Praise for
The Enduring Wilderness

"Holy, Holy, Holy!"

—Kurt Vonnegut

"An invaluable guidebook to saving America's wilderness. A must for ordinary citizens who care about saving our wilderness heritage for future generations. Hopeful, practical, and compelling."

—Christopher Reeve

"Tremendous! Doug Scott captures the heart, soul, and vision that created America's wilderness heritage. This book is definitely a call to action for all who desire to see wilderness preserved for future generations to experience and appreciate. Our wilderness lands are only safe if we take a stand. Doug has certainly articulated the case for wilderness in this book."

—Dennis Madsen
President and CEO
Recreational Equipment Inc.

"Here are the why and the how of ordinary people who use our democratic system to preserve wild places they love. One of the valued 'elders' of our work, Doug has distilled four decades of his own experience as well the best of our movement's historical experience and political insights. An essential volume for everyone who wants to help save an enduring resource of wilderness."

—William H. Meadows
President
The Wilderness Society

"*The Enduring Wilderness* is well-written, convincing the reader to leave it wild."

—Cindy Shogan
Alaska Wilderness League

The Enduring
Wilderness

DOUG SCOTT
Campaign for America's Wilderness

Fulcrum Publishing
Golden, Colorado

Library of Congress Cataloging-in-Publication Data Pending

ISBN 1-55591-527-2

Printed in the United States of America
0 9 8 7 6 5 4 3 2 1

Editorial: Kristen Foehner, Faith Marcovecchio, Katie Raymond
Design: Jack Lenzo
Cover image: © Wendy Shattil/Bob Rozinski

Cover: Kayakers enjoy the wild scene in the Brothers Islands, part of the 956,000-acre Kootznoowoo Wilderness within Admiralty Island National Monument, in the Tongass National Forest in southeast Alaska. Many other roadless portions of the Tongass are being advocated for wilderness designation in the face of continuing threats from subsidized logging. For further information visit www.seaac.org.

Fulcrum Publishing
16100 Table Mountain Parkway, Suite 300
Golden, Colorado 80403
(800) 992-2908 • (303) 277-1623
www.fulcrum-books.com

Dedication

For Kiana Melrose Scott and Kendall Selkirk Scott and all who work to protect more of our national heritage through the Wilderness Act.

In devoted memory of the one who, more than any other, gave us the tool to preserve wilderness with a presumption of perpetuity, Howard Zahniser (1906–1964).

Table of Contents

Foreword

Farsighted decisions taken by earlier generations of Americans have given our nation an extraordinary bequest of public lands that stretch from northern Alaska to southern Florida, amounting to almost 29 percent of the country's landmass. No other country has done anything remotely comparable. The story of preserving land for public use and for future generations began in 1872 with the establishment of Yellowstone National Park. It was a revolutionary idea. In Europe only the aristocracy had access to open space and the activities it provided. Some of our early conservationists, such as the artists Albert Bierstadt, George Catlin, and Thomas Cole, or writers including Walt Whitman and Henry David Thoreau, recognized that what made America different and had a profound impact on its people were its mountains, prairies, and wilderness.

As early as 1923, Colonel William B. Greeley, a thoughtful chief of the U.S. Forest Service, issued a plea that "stretches of untrammeled wilderness" be reserved where only a few of the more hardy and "elect" among the seekers of the out of doors "can penetrate relying upon their unaided skill in woodcraft." Greeley was responding to the desire of many Americans to preserve some land that would remain unchanged from the time it was first seen by Daniel Boone, Kit Carson, Jim Bridger, Meriwether Lewis, and William Clark and known only by the Native Americans.

A year later in 1924, led by Aldo Leopold the Forest Service established the first wilderness area. The idea of preserving wilderness areas, he wrote, "is premised on the assumption that the rocks and rills and templed hills of this America are something more than economic material." Wilderness, he said, "is the very stuff America is made of."[1]

Two politicians who understood and supported Leopold's desire for wilderness preservation were Clinton P. Anderson, who as a senator from New Mexico shepherded the Wilderness Act through the Senate along with his counterpart in the House of Representatives, John P. Saylor from Johnstown, Pennsylvania. Their efforts resulted in the Wilderness Act, which became the law of the land with President Johnson's signature in 1964.

Today, four decades later, the Wilderness Act is rightly ranked among the most important conservation achievements of this nation and serves as a model for other countries. Greeley did not foresee that our wilderness areas would become extraordinarily popular with backpackers of all ages, young families introducing their children to nature, and even retirees birdwatching on day hikes.

The essential roles played by Senator Anderson, an ardent Democrat, and

Representative Saylor, a stalwart Republican, remind us that conservation and wilderness have properly enjoyed bipartisan support.

In this book, Doug traces the history of conservation and wilderness protection and draws from that history the real lessons of wilderness politics. He is a great guide to this rough political terrain, having been a wilderness lobbyist, strategist, and historian. He began his career when he fell in love with Isle Royale National Park in Michigan. Doug's leadership, along with others', ultimately persuaded Congress to designate 98.7 percent of the park land as wilderness.

When I returned to the United States after my second tour in Vietnam, I immediately went camping in the High Sierras. At the time, I told myself that I wanted a contrast with the hot lowland jungle where I had spent the prior months. That was partly true, but I was also seeking solace and spiritual renewal. The need for wilderness and the refuge it offers to Americans will only increase with the passage of time. Our generation of Americans has an obligation to preserve for future generations more areas that qualify for wilderness designation. This designation provides the strongest form of protection and requires an act of Congress that can only be altered by an act of Congress. Theodore Roosevelt once said, "of all the questions which can come before this nation, short of the actual preservation of its existence in a great war, there is none which compares in importance with the great central task of leaving this land even a better land than it is for us. ... "[2]

—Theodore Roosevelt IV

Preface

From space Earth is a tiny ball, blue-green with the colors of water and life. As our view zooms in on North America and the lower forty-eight states, the widespread impacts of human development are staggering. *Seven million miles of roads!* Yet nearly 5 percent of all the land in the United States remains wilderness, securely protected by federal law—the 1964 Wilderness Act—with more to come.

That so much American wilderness has been preserved by law, and more will be, is the work of far-sighted senators, representatives, and presidents of both major political parties. The Wilderness Act—and later laws extending its protections to additional federal lands—was accomplished with the devoted assistance of many public servants, whom we entrust with the careful stewardship of our wilderness areas. All of this success was propelled by unstoppably persistent citizen-advocates. Foremost was Howard Zahniser, the father of the Wilderness Act, whose life's work is at the heart of much of this story and who has been my mentor, though I never knew him.

The wilderness preservation movement is larger today than ever before. Gifted and deeply dedicated volunteers and staff are engaged in this work in every state, backed by large and robust national organizations. It has been my life's blessing to work with so many who have done so much for wilderness. My purpose here is to inspire more such spirited advocacy for wilderness.

The human impulse to protect some of our once all-wilderness planet, leaving some places where nature may unfold in its own way, is all about our obligation to generations to come. Beyond all the other good reasons for saving wilderness, we do so because it would be morally inexcusable not to be wildly generous with future generations.

And consider this, though it may be hard to imagine: should our generation overdo it, preserving "too much" wilderness, those who come after will smile on us for our good grace in leaving the choice to them. They may always correct such an "error," should they wish.

But here is what I think: *however much wilderness we in our own time choose to preserve by law, future generations will be more likely to judge that we protected too little than that we protected too much.*

Today, we look backward to a time when there was more wilderness than the people of America needed. Today we look forward (and only a matter of a few years) to a time when *all* the wilderness now existing will not be enough.

—William O. Douglas, foreword,
The Wild Cascades: Forgotten Parklands

xi

To give our grandchildren's grandchildren—and theirs—this wilderness choice should be our gift, and small enough recompense it will be for all that we have done to, and taken from, so much of *their* land.

This sense of intergenerational obligation to work for a wilderness-forever future came alive for me with the birth my daughters, Kiana and Kendall. In the choices their generation will make about wilderness, they have my loving proxy.

Howard Zahniser, executive director of The Wilderness Society, 1945–1964, drafted the Wilderness Act and led the campaign to make it law. Publicity photo used by The Wilderness Society about 1959. (Courtesy of The Wilderness Society)

Introduction

People Saving Wilderness ... for People

In the 1964 Wilderness Act, Congress declared a national policy to "secure for the American people of present and future generations the benefits of an enduring resource of wilderness." Congress concluded that only lands deliberately set aside with firm legal protection could reasonably be expected to persist in their natural, undeveloped, roadless condition "for the use and enjoyment of the American people." The Act established the National Wilderness Preservation System and designated the first fifty-four wilderness areas—9 million acres of federal land—protected as wilderness by law.

Applying the Wilderness Act, Congress has subsequently enacted more than 100 additional laws adding more lands to the wilderness system. Each wilderness area gains *statutory* protection within boundaries fixed by Congress. This affords the strongest possible security, for only Congress can then alter those boundaries or the degree of protection given to those lands. The basics of the program set out in the Wilderness Act are straightforward:

- The lands protected as wilderness are areas of our *public lands*, owned in common by the American people. These federal lands are found in all regions of the country, in national forests, national parks and monuments, national wildlife refuges, and on other public lands.
- Four federal agencies manage wilderness areas on lands within their jurisdiction— the U.S. Forest Service (in the Department of Agriculture) and the National Park Service, Fish and Wildlife Service, and Bureau of Land Management (all in the Department of the Interior). Wilderness designation is a *protective overlay* Congress applies to selected portions of these differing categories of public lands. Roads, logging, oil and gas drilling, and other development are allowed on some public lands—but not in the portions designated as wilderness.
- Within wilderness areas, we strive to restrain human influences so that ecosystems can change over time in their own way, as much as possible, *free from human manipulation*. In these areas, as the Wilderness Act puts it, "the earth and its community of life are untrammeled by man"—untrammeled meaning unrestrained and unaltered.
- Wilderness areas *serve multiple uses* and provide many benefits. But the law limits uses to those consistent with the Wilderness Act mandate that each wilderness area be administered to preserve the "wilderness character of the area." For example,

1

these areas protect watersheds and clean-water supplies vital to downstream municipalities and agriculture, as well as habitats supporting diverse wildlife, including endangered species, while logging and oil and gas drilling are prohibited.

- Along with many other uses and values for the American people, wilderness areas are popular for diverse kinds of outdoor recreation—but *without motorized or mechanical vehicles or equipment*. Wilderness is the haven of quiet beyond the end of the road, the wild sanctuary we meet on its own terms by leaving the machinery of twenty-first-century life behind. The wild popularity of wilderness recreation shows how hungry Americans are for just such sanctuaries.

The extent of the nationwide system of wilderness areas Congress has built using the Wilderness Act reflects a strong national consensus to save a generous sample of our undeveloped federal lands. By mid-2004 Congress had protected nearly 106 million acres in the wilderness system in more than 650 areas in forty-four of the fifty states. More than half of the total expanse is in Alaska. Outside Alaska, wilderness protections apply to just 2.5 percent of all the land in our country—not just public land, but land in all ownerships.

This is not all the wilderness that Americans will choose to preserve on their public lands. Congress considers additional proposals every year, some recommended by federal agencies and many proposed by citizens and grassroots conservation and sportsmen's organizations.

As this book went to press in the summer of 2004, bills to designate millions of acres of new wilderness areas were pending in Congress involving federal lands in Alaska, California, Colorado, New Mexico, Nevada, Puerto Rico, Utah, Virginia, and Washington. Grassroots coalitions are working with local congressional delegations on legislative proposals for additional wilderness areas in the sageland desert and the mountains of Idaho; the rich forests of Vermont and New Hampshire; the desert hills of southern Arizona; the grasslands of South Dakota and Wyoming; the Rocky Mountain peaks of Montana; the piney woods of the South; the forests of Oregon and Washington; and the north woods of Minnesota, Michigan, and Wisconsin. The U.S. Forest Service has recommended new wilderness designations (which citizen groups may propose to expand) in some of these places and others.

In fact, wherever there are public lands with fragile wilderness qualities, you will find citizen groups working to secure the strongest protection we can give such lands, which is wilderness protection by act of Congress.

The idea within the word "Untrammeled" of [wilderness areas] not being subjected to human controls and manipulations that hamper the free play of natural forces is the distinctive one that seems to make this word the most suitable one for its purpose within the Wilderness Bill.

—Howard Zahniser to C. Edwards Graves, April 25, 1959

The Impulse to Preserve Wildness

Before recorded history and into recorded time for American Indians and other indigenous peoples, humans lived as part of the natural world. They were of the wilderness. But as human societies evolved, for many wilderness came to be distinct— a fearsome reality of wild beasts and brigands beyond the tenuous protections of settlement and stockade. Yet, eventually under the influence of artists, poets, and philosophers, Western culture evolved to reverse the ancient fear of wild nature. Euro-American culture came to value and treasure wilderness. As the American frontier was pushed back and the tide of settlement tamed much of the wilds, more and more Americans wished to avoid seeing wilderness vanish from our landscape and our culture. Wilderness, Aldo Leopold wrote in 1925, "is the very stuff America is made of."[1] Many of us work to protect wilderness areas—to assure that they are part of America's future as well as its past—to preserve a touchstone to the untrammeled natural world. But wilderness is a living reminder, too, of the deep impress that our encounter with the wild frontier has given to our history and our national character as a people.

This book chronicles the story of how, in the first half of the twentieth century, pioneering wilderness advocates in and out of government struggled to find a practical way to actually protect some areas of wilderness. As they came to see it, wilderness was not some luxury or plaything but a fundamental human *need*, yet the American wilderness was, as they expressed it, "vanishing with appalling rapidity." Their goal was nothing less than to preserve wilderness areas with "a presumption of perpetuity." As these preservation advocates came to discover that other approaches could not give that guarantee of permanence, they concluded by midcentury that wilderness preservation could be secured only by statutory law—by act of Congress.[2]

It took eight years of often contentious debate for Congress to enact the Wilderness Act in 1964. Most of the debate was about details, not the act's goal. All involved well appreciated that this law was destined to be among our most historic conservation policies. Leading the advocacy for the bill on behalf of President John F. Kennedy, Secretary of the Interior Stewart Udall told senators that the bill ranked with the Homestead Act of 1862 and the National Parks Act of 1916 as a "great landmark conservation decision."[3]

This sense of the act's fundamental importance and the strong congressional support for the law were thoroughly bipartisan. In a special message ten years after the Wilderness Act became law, President Gerald R. Ford told Congress that preserving wilderness "serves a basic need of all Americans, even those who may never visit a wilderness area—the preservation of a vital element of our heritage."[4]

In 2000 a survey of history and political science scholars ranked the Wilderness Act and its implementation as twenty-fifth among the fifty most important accomplishments of the federal government in the past half century, rating it just below arms control and just above space exploration.[5] And thanks to this law, a goodly sample of America's wilderness lives on.

The Wilderness Act: A People's Law

When President Lyndon Johnson signed the Wilderness Act on September 3, 1964, it immediately protected by statute just 9 million acres of wilderness. Nearly eleven times as much—97 million acres—has since been added to the wilderness system, reflecting the hard work of legislators of both political parties, citizen-advocates, and agency personnel.

The Wilderness Act is a people's law. As former Senators Dale Bumpers, a Democrat, and Dan Evans, a Republican—both champions of wilderness designations in their own states, Arkansas and Washington, respectively—have written

Protecting wilderness areas is not some top-down federal decision. It is the most "small-d" democratic land allocation process we've invented. Potential wilderness areas are identified by on-the-ground agency staff and local people who know the land best, and then the decision to carry a bill is made by our elected representatives in Congress, led usually, though not always, by the local congressional delegation.[6]

In an age of automation, mechanization, and exploitation of our vast natural resources, the amount of public lands shielded from the onslaught of man's ambition and genius becomes [ever] smaller. Our task in this age has been to stand off and ponder the consequences of that onslaught. I believe that this bill contains our verdict, and I believe that we can all be grateful that the verdict came while we still had wilderness to preserve.

—Senator Clinton P. Anderson,
Congressional Record, August 20, 1964

Wilderness laws have been the good works of deep-dyed conservative Republicans and Teddy Roosevelt progressive Republicans, of "Blue Dog" Democrats and liberal Democrats, and of Independents. Since President John F. Kennedy personally advocated the bill and President Lyndon B. Johnson signed it into law, every president has contributed by signing further legislation designating additions to the wilderness system. In the current political metaphor, wilderness has been protected in both "blue states" and "red states."

That so many areas across America have gained wilderness protection is, most of all, testament to the dedication of local people who have organized and persevered for years, building the constituent support that can ultimately fuel action by their members of Congress. And this work goes on today all over the country.

- When road building threatened to invade a favorite wild place in the mountains behind the small town of Lincoln, Montana, Cecil Garland, owner of the hardware store, mobilized grassroots support that led to the 1972 law protecting the quarter-million-acre Scapegoat Wilderness, part of Montana's Bob Marshall wilderness complex.
- When logging was proposed for mountain slopes she loved, Katherine Petterson, a

pear grower in Kelseyville, California, became a leader of the campaign that persuaded her congressman and ultimately the entire Congress to enact statutory protection for the 36,000-acre Snow Mountain Wilderness, achieved in 1984.

- To protect an area of rare and endangered species' habitat where she likes to hike, Nancy Hall, a waitress in a Nevada casino, helped organize the campaign that preserved the 23,000-acre Lime Canyon Wilderness, signed into law in 2002. Since then she has been writing letters and circulating petitions to protect additional areas. "I guess I'm an activist," she told the *Las Vegas Sun*. "It took me a long time to say that's what I am."[7]

Sometimes, these achievements have been made in close collaboration between citizen-activists and the federal agency administering the particular lands. In other cases, while local agency personnel might be sympathetic and helpful, their Washington-level superiors or an unsympathetic presidential administration may overrule them and oppose the wilderness proposal. But, thanks to the Wilderness Act, administrative agencies no longer have the final word. Citizens can take their maps, field studies, and proposals directly to Congress. Some agency officials have found that this offends their sense of their own control of land-use decisions. In the case of Cecil Garland's citizen proposal for the Scapegoat Wilderness, the regional head of the U.S. Forest Service was affronted by the greater influence the Wilderness Act gives citizen-advocates who take their case to their members of Congress:

> Why should a sporting goods and hardware dealer in Lincoln, Montana, designate the boundaries for the 240,000-acre [Scapegoat Wilderness]? If lines are to be drawn, we should be drawing them.[8]

But the rules had changed. Congress had decided that it alone would draw the lines and that the full strength of statutory law was necessary to achieve its goal—to protect wilderness, in the words of the Wilderness Act, "for the permanent good of the whole people."

So, the story of our National Wilderness Preservation System is one with many players—first artists, poets, and philosophers, then pioneering leaders trying to find a practical way to preserve areas of wilderness, then lobbyists and members of Congress of both political parties who fought to enact the Wilderness Act, then agency personnel entrusted to be stewards of these wild places. It is the story, too, of the members of Congress, federal agencies, nonprofit organizations, and tens of thousands of citizens who have worked and are

> We are trying to keep unchanged by man areas that have so grown through the eternity of the past, and, although we stand in awe at our own presumption, we dare to plan that they may so persist through the eternity yet ahead. ...
>
> Hope in the United States for wilderness in the future depends on our success in developing a policy and program that provide for the preservation for wilderness as such, by our Federal government, with a presumption of perpetuity.
>
> —Howard Zahniser, "The Wilderness Bill and Foresters," March 14, 1957

working now to add areas to the system, thereby protecting them with all the strength of statutory law.

Challenges Ahead

The National Wilderness Preservation System is far from complete, whatever "complete" might mean. There is not some advance master plan or arbitrary cut-off point defining what will be "enough" wilderness. Nor can there or should there be. The choice of which areas of wilderness will be protected on the people's own public lands are made, in this most "lowercase-d" democratic of all land-use decisions, by those the people elect to represent them in Congress.

This book recounts not only how America's wilderness program came to be and came to protect nearly 106 million acres, but how additional areas will be added to our National Wilderness Preservation System. For many reasons, the politics of wilderness designation are not static.

- The most forceful opposition to wilderness designation varies. In some places at some times it may be logging companies or mining interests that oppose statutory protection of wilderness. In recent years, oil and gas drilling companies have taken a larger role in many places.
- The arguments for and against wilderness preservation evolve and change, even as the larger politics of our nation evolve and change. Progress in designating additional wilderness areas depends, in some cases, on such factors as which party controls the White House, the Senate, and the House of Representatives, as well as on the swings in the politics of each state.
- Technical issues arise concerning aspects of some wilderness proposals, are resolved, yet tend to rise again for fresh debate. Among these issues are
  - What degree of past human disturbance can have occurred in a candidate area?
  - How do we deal with established but nonconforming uses such as livestock grazing?
  - What is the role of potential wilderness areas in the context of uses of the surrounding lands?
- Everywhere, the appetite of some off-road vehicle users to drive wherever they please across public lands has become a pervasive threat to wild lands.

Many substantially undeveloped, roadless areas amounting to tens of millions of acres remain on our federal lands. Much of this land will not be designated as wilderness, but all of it merits serious review. As the U.S. Forest Service and the Bureau of Land Management plan for the use of lands in their jurisdictions, agency officials should fairly consider the option of seeking congressional designation for lands that are wilderness in fact but not yet in law. This *unprotected wilderness* is the focus of discussion in many states.

Designating wilderness areas is only the first step. Then comes the great challenge of stewardship. I try to avoid the oxymoron *wilderness management*. Of course, the personnel of the four federal wilderness agencies—the Bureau of Land Management, U.S. Forest Service, Fish and Wildlife Service, and the National Park Service—are land managers. Many of their personnel were trained, as I was, in forestry or in other land-management specialties. But where wilderness is concerned, we must strive for a different mind-set. The better term *wilderness stewardship* underscores that the essence of our task is humility, the instinct to withhold our manipulative urges. For wilderness areas, in Howard Zahniser's perfect epigram, we should strive to be "guardians, not gardeners."

Public opinion polling confirms very strong, very broad public support for wilderness protection, regardless of region, across the urban/rural divide, and across party lines. Inevitably there will be efforts to open wilderness areas for development. But to do so will require Congress to pass a new law.

In the final analysis, wilderness is preserved not by a law engrossed on parchment but by the people themselves. Because the American people so strongly value these wild places, sharing a sense of duty to pass them unimpaired to future generations, we have a strong wilderness program backed by a strong law. Because Americans have an inborn love of their land—of their "purple mountain majesties"—we can have confidence that wilderness will remain a part of America's long future.

To the pioneer of history the wilderness was a foe to be conquered, so that he might make farms and pastures out of the endless forests.

Today's pioneer has a new purpose— to preserve some remnants of that wilderness from the onrush of modern civilization.

The ax and the plow will not serve us in this struggle. Today's instruments are more subtle. They are progressive law and informed public opinion—demanding that we maintain our wilderness birthright.

—President Lyndon B. Johnson
on the Status of the National Wilderness
Preservation System, February 14, 1966

7

Chapter 1

Wilderness in Our Lives

The Wilderness We Have Saved

The statistics are impressive. In the first forty years under the Wilderness Act, Congress has enacted laws protecting 4.7 percent of all the land in the United States. This comprises some 106 million acres. But what does that really mean?

Our wilderness areas range from a single unbroken expanse of nearly 13 million acres centered in Alaska's Brooks Range to a 3,700-acre wilderness area in suburban New Jersey, just twenty-six miles from Times Square. Our smallest wilderness area is a five-acre island in Florida; among our most "urban" are wilderness areas literally on the city limits of Albuquerque; Logan, Utah; and Tucson.[1] As Congress passed Senator Pete Domenici's bill to expand the Sandia Mountain Wilderness on the city limits of Albuquerque, the senator, a New Mexico Republican, told his colleagues that the area "forms a beautiful natural backdrop for the city which all the residents can enjoy." Senator Domenici was highlighting a use and benefit of wilderness too often neglected by those who assume that the scenic values of wilderness are only of benefit to those *within* its boundaries.[2]

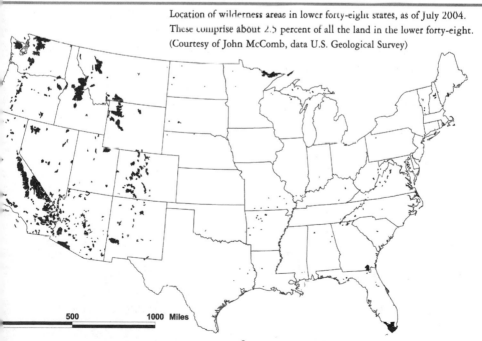

Location of wilderness areas in lower forty-eight states, as of July 2004. These comprise about 2.5 percent of all the land in the lower forty-eight. (Courtesy of John McComb, data U.S. Geological Survey)

500   1000 **Miles**

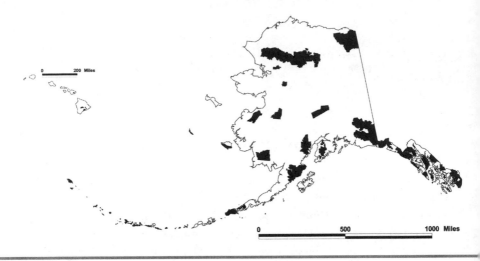

Wilderness areas in Hawaii and Alaska (which has over half of the total acreage in the
wilderness system) as of July 2004. (Courtesy of John A. McComb, data U.S. Geological Survey)

These and our other wilderness areas have been protected not as part of some
grand master plan, but in an incremental process of Congress responding to
constituent interest. Often Congress has later expanded initial wilderness
designations in response to public interest. Consider one example. By administrative
order in 1931, the U.S. Forest Service protected the 55,000-acre Ventana Primitive
Area in the central California coastal mountains above the Big Sur coast. After a
study required by the Wilderness Act, the area was expanded to 98,000 acres and
given statutory protection by Congress in 1969 as the Ventana Wilderness. Congress
has since returned to this area four times, passing laws adding increments of
continuous lands in 1978, 1984, 1992, and 2002. Today, the Ventana Wilderness
protects almost a quarter-million acres.

Just as our National Park
System has grown by establishment
and expansion of parks over more than
130 years, so the work of preserving
wilderness by law—which has only
been underway for four decades—
remains a work in progress. "It is
hard," write John Hendee and Chad
Dawson in the *International Journal of
Wilderness*, "to identify any natural
resource issue—or any issue—for
which Congress has so consistently
and so often confirmed their intent
as they have with wilderness."[3]

The National Wilderness Preservation System: Acres Administered by Each Federal Agency (July 2004)		
Agency	Wilderness acreage	Percent of agency land designated wilderness
National Park Service	43,616,250	56%
U.S. Forest Service	34,867,591	18%
U.S. Fish and Wildlife Service	20,699,108	22%
Bureau of Land Management	6,512,227	2%
Total	105,695,176	17%

DATA SOURCES

Wilderness acreages by agency from www.wilderness.net. For consistency, all
acreage data used for percentage calculation are from *Federal Land Management
Agencies: Background on Land and Resource Management* (Washington, D.C.:
Congressional Research Service, RL30867, February 2001): 7.

We can say some things to help characterize the wilderness areas now protected in the National Wilderness Preservation System:

- Areas in our wilderness system are very *diverse.*
  - Wilderness areas are diverse in *geography and ecosystems*, embracing freshwater swamps and coastal salt marshes, glaciers and rocky mountain peaks, deeply forested valleys and ridges, broad desert plateaus and canyons, vast sweeps of Arctic tundra, tumbling mountain brooks and great rivers.
  - Wilderness areas are diverse in *size.* There is no minimum; Congress just makes a judgment regarding what size is practical in order to preserve even a small area in perpetuity.
  - Wilderness areas are diverse in their *context and settings.* No one would argue that a small wilderness swamp in New Jersey is the same thing as a multimillion-acre wilderness in Alaska. In its own setting each protects real wilderness values. For millions of Americans, wilderness areas are not only faraway locales for once-in-a-lifetime adventure, they are also close and familiar old friends.
  - Wilderness areas are diverse in *degree of wildness.* Some wilderness areas are as close to pristine as is possible in a world so heavily impacted by human development. Others have a history of past development and human impact but are recovering under natural forces.

Just as a sample, let me suggest the Robbers' Roost country in ... Utah. It is a lovely and terrible wilderness, such a wilderness as Christ and the prophets went into; harshly and beautifully colored, broken and worn until its bones are exposed, its great sky without a smudge of taint from Technocracy, and in hidden corners and pockets under its cliff the sudden poetry of springs. Save a piece of country like that intact, and it does not matter in the slightest that only a few people every year will go into it. That is precisely its value. Roads would be a desecration, crowds would ruin it. But those who haven't the strength or youth to go into it and live can simply sit and look. They can look 200 miles, clear into Colorado; and looking down over the cliffs and canyons of the San Rafael Swell and the Robbers' Roost they can also look as deeply into themselves as anywhere I know. And if they can't even get to the places on the Aquarius Plateau where the present roads will carry them, they can simply contemplate the idea, take pleasure in the fact that such a timeless and uncontrolled part of the earth is still there. ...

We simply need that wild country available to us, even if we never do more than drive to its edge and look in. For it can be a means of reassuring ourselves of our sanity as creatures, a part of the geography of hope.

—Wallace Stegner,
"Wilderness Letter" to the Outdoor Recreation Resources Review Commission,
December 3, 1960

This great diversity of our wilderness areas reflects the natural diversity of our landscapes and reflects, too, the unique land-use history of each area and the sparseness or density of surrounding human development. In a 1949 essay Aldo Leopold wrote that "wilderness is the raw material out of which man has hammered the artifact called civilization" but that "many of the diverse wildernesses out of which we have hammered America are already gone; hence in any practical program the unit areas to be preserved must vary greatly in size and in degree of wildness."[4]

- Areas in our wilderness system provide *many benefits and uses*.

Wilderness areas serve multiple uses, providing many benefits for all of us. It is too commonly assumed that wilderness is primarily about recreation. Of course, wilderness recreation is very popular—for backpacking, day hikes, birding, youth outings, camping, wilderness-quality hunting (in most wilderness areas other than those within national parks), fishing, nature study, horseback riding and packing (including with commercial guides and outfitters), canoeing and kayaking, wildlife and wildflower viewing, cross-country and backcountry skiing, snowshoeing, and wilderness photography, all in an environment of solitude and quiet. But wilderness serves many other uses as well:

- Wilderness areas protect watersheds, providing abundant *clean water* for millions of Americans who live downstream.
- They are reservoirs of *clean air*, protecting and promoting pristine air quality.[5]
- They are a place for *wildlife* to flourish, protecting flora and fauna and their habitats, including many threatened wilderness-dependent species such as grizzly bears.
- Wilderness areas protect *rare and endangered species* and their habitats, including many endangered plants and spawning beds for endangered species of fish, such as wild salmon.
- They protect the integrity of views, allowing roadside enjoyment of sweeping *wilderness scenery* for residents and visitors alike, a living reminder of the vastness of wilderness frontiers of yesteryear.
- Protecting wilderness areas serves deep *cultural and historical purposes*, preserving a living reminder of the pioneer encounter with the wilderness that so fundamentally shaped American culture and national character.
- Wilderness areas speak to our *spiritual values*. One of the architects of the Wilderness Act said it this simply: wilderness areas are "buffer areas for the human spirit."[6]
- They allow us to protect our *options*, preserving areas that we may choose to visit in the future.
- Wilderness areas protect a *legacy* of the original American earth, which so many Americans feel it is the duty of our generation to pass on, undefiled, as the natural heritage we owe to future generations.

The recreational values so strongly identified with wilderness preservation have in common a special and unique quality. In the words of Congressman John Saylor, the champion for the Wilderness Act in the House of Representatives, encouraging enjoyment of wilderness is about

> putting the American citizen back on his own two, God-given feet again, in touch with nature. Let's get him away for a blessed few moments from the omnipresent automobile, if we can, rather than catering to what is easiest and least exerting. And let's certainly not turn every quiet wilderness pathway into a "nature trail" for motors.[7]

What Preserving Wilderness Requires

What does it mean to "preserve" or to "designate" a wilderness area? Some elements of wilderness preservation are fundamental:

Even if a wild place seems relatively safe from development, threats are ever present and new threats will emerge.

To the Congress:

Inevitably, our work as public leaders affects not only us but our posterity. It may be that our most important constituents are not our contemporaries, but the men and women who will inherit America from us in 20 years or 50 years.

Our grandchildren and our great-grandchildren will live in a different America from the one we know; but it will be an America we have helped to build. They will work in buildings and factories we erect; travel on highways we lay out; live in cities and suburbs we construct; seek solace and recreation in parks and wilderness areas we preserve.

So we must build well now.

Unless we do, much of the wild and beautiful America that we know, and our grandfathers knew, will be lost forever—buried in the debris of our hurrying civilization.

Many of these [wilderness] areas ... are close to the centers of American population. They can and will be enjoyed by millions of our people seeking the solitude and splendor of the land as God made it. So they are a trust and a responsibility for all of us.

—President Lyndon B. Johnson,
letter to Congress, March 29, 1968

The 7 million miles of roads in the lower forty-eight states make a map virtually black over much of the country. For these states in the Northwest, the existing wilderness areas and other still-roadless federal lands stand out like islands of quiet in that human-dominated landscape. View the full map at www.pacificbio.org. Pacific Biodiversity Institute sells a stunning 36" by 48" print of their famous road map; see the Web site for details. (Courtesy of Peter Morrison)

The impacts of human development are pervasive in our twenty-first century world. Wild places cannot protect themselves, for the pressures to invade them is intense and never ending. Even in 1925 Aldo Leopold was warning of these pressures:

> Do not forget that the good roads mania, and all forms of unthinking Boosterism that go with it, constitute a steam roller the likes of which has seldom been seen in the history of mankind. ... It might very readily flatten out, one by one, the remaining opportunities for [preserving wilderness].[8]

On a continent that was all wilderness not so long ago, the western course of empire has imposed man's will over the vast majority of the landscape. Wild forests have been cut down and roads pushed up nearly every valley and canyon and across virtually every plateau. The impacts of roads and all they bring are the very death of wilderness—and in little more than a century, we have built 7 *million* miles of roads in the lower forty-eight states! Roads fragment and refragment large, intact ecosystems into ever smaller, more unsustainable bits and pieces, destroying opportunities for solitude and driving off wilderness-dependent wildlife. The mushrooming of ill-policed off-road vehicle use is spreading informal "roads" even farther.

In the face of such threats, the essence of preserving wilderness is to make an unequivocal decision that designates wild lands for the full protections of the Wilderness Act.

The effectiveness of wilderness protections relies upon establishing firm boundaries.

The imperative in preserving wilderness is to draw lines—and to draw them firmly. Unless boundaries are established around each wilderness area, one new development after another nibbles away at wild places in an insatiable, creeping process fatal to wilderness.

Responding to proposals to set wilderness boundaries far back from the edge of roads and other current development, Senator Frank Church, an Idaho Democrat who was a key leader in enactment of the Wilderness Act, lectured the agencies at a 1972 Senate hearing, saying

> There is no requirement for that in the Wilderness Act ... [such a boundary] has the effect of excluding the critical edge of wilderness from full statutory protection ... In the absence of good and substantial reasons to the contrary ... the boundaries of wilderness areas ... should embrace all wild land. There is no lawful policy basis for massive exclusions of qualified lands on which no development is planned.[9]

Thus a crucial condition for truly preserving wilderness areas is to set boundaries that protect not only the core of the area but also what Senator Church called "the critical edge of wilderness," precisely to stop increments of development from nibbling away at wild land everywhere.

History demonstrates that administrative *protection for wild lands simply cannot assure firm boundaries in the long run.*

Development interests have great economic and political influence. Whether timber companies, oil and gas developers, or off-road vehicle groups, they wield their clout to press for yet another timber sale, oil and gas lease, or extension of motors, each probing ever farther into undeveloped wild lands. To wilderness leaders in the decades before the Wilderness Act, it was painfully clear that relying on agency administrative decisions to draw a firm boundary line and stick by it when under this kind of unrelenting development pressure was simply unrealistic. An administrative order adopted by the stroke of a pen can be just as easily altered or revoked.

In fact, in the decades before the Wilderness Act, administratively established boundaries of "protected" wilderness lands were repeatedly revised, almost invariably to release areas for logging and other development. Even within national parks, Congressman John P. Saylor concluded, "administrators ... beset by pressures for inappropriate development and overdevelopment, never have set a firm and final limit on the extent of development that is permissible if the preservation of the primeval parks is to be secure."[10]

Having had to rise to the defense of such places time after time, in the 1950s wilderness leaders turned to the idea of a wilderness law so that boundaries would be far less easily revised whenever politics or economic pressures could prevail. Democratic Senator Richard L. Neuberger of Oregon, an early champion of the Wilderness Act, said of the proposed law in 1957

> Wildlife, waterfowl, migratory fisheries, and similar resources require outdoor fastnesses and solitudes in which to survive, and ... these must be safeguarded by some form of legislative shield.[11]

In 1956, as he introduced the first wilderness bill that led to the 1964 act, Congressman Saylor told his colleagues in the House of Representatives that "our only democratic hope for success in preserving our wilderness resource is in our policy of deliberately setting aside such areas for preservation and then maintaining the integrity of our designation." Looking back on that fundamental motive for the Wilderness Act, Saylor later recalled

> Even at that time we saw that ... wilderness needed to be designated positively and securely, so that it did not end up simply being whatever residue of wild land remained after each further stage of development.[12]

We have no difficulty in recognizing the general purpose of the Wilderness Act. It is simply a congressional acknowledgment of the necessity of preserving one factor of our natural environment from the progressive, destructive and hasty inroads of man, usually commercial in nature, and the enactment of a "proceed slowly" order until it can be determined wherein the balance between proper multiple uses of the wilderness lies and the most desirable and highest use established for the present and future. A concerned Congress, reflecting the wishes of a concerned public, did by statutory definition choose terminology that would seem to indicate its ultimate mandate.

—Parker v. United States, 448 F.2d 795,
U.S. Court of Appeals, Tenth Circuit

Only protection by statutory law—by act of Congress—can reasonably guarantee protection of wilderness in perpetuity.

Perpetuity is the very essence of wilderness preservation. Nothing less is good enough. We save these special places not for ourselves alone but most of all for generations unborn. Therefore, nothing less than the strongest possible form of security can be accepted.

As early as 1939, experience with piecemeal nibbling away of the wilderness lands within national forests and national parks led national conservation leaders to an inescapable conclusion, voiced by Kenneth Reid, head of the Izaak Walton League, in that organization's magazine:

> There is no assurance that any one of them [wilderness areas], or all of them, might not be abolished as they were created—by administrative decree. They exist by sufferance and administrative policy—not by law.[13]

So, in the middle of the twentieth century, national wilderness leaders made the fundamental choice to turn to Congress to seek a law that would provide for the fixing of firm boundary lines, backed by the full force of statutory law.

Why is statutory protection the best guarantee for permanent wilderness designations and boundaries? It comes down to this: acts of Congress are not easy to enact. By design, the writers of the Constitution set up a bicameral legislative branch, requiring any law to be passed through both the House and Senate in identical form and then be signed by the president. The procedures were planned by visionaries who had thrown off the yoke of a king and the arbitrary imposition of laws by parliament; they intended enactment of laws to be a difficult procedure which tilted the burden of proof, giving opponents plentiful opportunities to slow the process or block approval.

Thus it is hard to enact a law to designate a new wilderness area. But the effort is worth it, for once the area is designated and its boundaries fixed by law, it takes another act of Congress to weaken protections or alter the boundary. Congress has enacted laws that removed some lands from designated wilderness areas about ten

---

I remember the beginnings of that one—the eight-year-long effort to pass a wilderness bill—the council of The Wilderness Society sitting on the shores of Rainy Lake in the Minnesota canoe country, facing the fact that administrative regulations were not enough to really protect wilderness, that we would be confronted with endless controversy, as one area after another would be threatened, that we needed a law to apply to all wilderness—forest, parks, refuges.

—Mardy Murie, *Arctic Dance*

times, each by very small increments to correct errors and each endorsed by wilderness groups. Had local or national wilderness groups been opposed, those laws would very likely have been blocked. President George W. Bush signed an act of Congress in 2003 to correct a boundary problem in a Utah wilderness area designated a decade earlier. That adjustment—by just thirty-one acres—required an act of Congress, exemplifying that the Wilderness Act gives our legislative branch the sole authority to draw or change wilderness boundaries.[14]

Opponents of additional wilderness designations understand the power of this statutory protection. Already, in the first four decades under the Wilderness Act, there have been periods when an unsympathetic administration has discouraged further wilderness designations. There have been, and will again be, political appointees who have an ax to grind on these issues, who may promise to administratively protect values associated with the wilderness concept but who oppose statutory protection.

Proponents of protecting wilderness began thinking of statutory protection long before it was possible to enact a wilderness law. Many of them were government officials, including the pioneering wilderness advocate Bob Marshall, who zeroed in on the fundamental rationale for a wilderness law in a 1934 memorandum:

> Under the present system a single unsympathetic administration could at any time wipe out our remaining primitive expanses. In order to escape the whim of politics, which might make the president of the American Automobile Association the next Secretary of the Interior and Henry Ford the next Secretary of Agriculture, the areas ... should be set aside by Act of Congress, just as national parks are today set aside. This would give them as close an approximation to permanence as could be realized in a world of shifting desires.[15]

Chapter 2

Wilderness: From Concept to Action

PART I: THE WILDERNESS IDEA

The Need for Wilderness

The founders of the wilderness preservation movement shared a passionate conviction that wilderness is not some luxury but a vital bulwark in our individual lives and the bedrock of our distinctive American culture. Their breakthrough concept was to link the existence of ecological wholeness most perfectly exemplified in wilderness areas to human well-being and our very civilization.

These gifted thinkers saw a fundamental human need for wilderness, not principally for recreational use or scientific study but for sustaining the American character shaped by our pioneer encounter with the wild frontier. Aldo Leopold spelled it out in 1925:

> In all the category of outdoor vocations and outdoor sports there is not one, save only the tilling of the soil, that bends and molds the human character like wilderness travel. Shall this fundamental instrument for building citizens be allowed to disappear from America, simply because we lack the vision to see its value?[1]

In founding The Wilderness Society in 1935, Leopold and his colleagues asserted that wilderness is "a natural mental resource having the same basic relationship to man's ultimate thought and culture as coal, timber, and other physical resources have to his material needs."[2]

This was something new, an American concept arising from our historical encounter with an all-wilderness continent. It was novel to see the pristine natural world not merely as a quarry of raw materials but as the very fabric of a distinctive American culture. To see in the reality of living wilderness "a fundamental instrument for building citizens" gave the wilderness preservation movement, from its very birth, a profound and energizing sense of mission.

The intellectual evolution of the wilderness idea in Western culture and in America's history and national ideology is traced by Roderick Nash in his indispensable study *Wilderness and the American Mind*. As Professor Nash notes, for eons as man became more and more dominant, "nature lost its significance as something to which people belonged and became an adversary, a target, merely an object for exploitation. Uncontrolled nature became wilderness," and wilderness—

literally the place of wild beasts—was feared.[3]

As we now appreciate, long before the Romantic era and the Enlightenment led to changes in this long hostility to wildness, hunting and gathering peoples would have found this distinction between wilderness and human society meaningless. Professor Nash quotes Chief Standing Bear of the Ogalala Sioux explaining that by tradition his people

> did not think of the great open plains, the beautiful rolling hills and the winding streams with their tangled growth as "wild." Only to the white man was nature a "wilderness" and ... the land "infested" with "wild animals" and "savage" people.[4]

The new idea of wilderness as deeply valued and fundamentally needed did not emerge all at once. It built on a long tradition in Western culture, on the thinking of Rousseau and Ruskin, the poetry of Wordsworth and Byron, and the Romantic-era fascination with wild natural spectacle evidenced in the Grand Tours to the Alps and the art of landscape painters. In the United States this new force in Western culture was expressed in the popular writings of Emerson, Thoreau, and George Perkins Marsh; in the narratives of naturalists William Bartram and John James Audubon; and in the novels of James Fennimore Cooper. It was depicted in the new idiom of the sublimity of wild nature, in evocative imagery of poets William Cullen Bryant and Walt Whitman, and in the works of artists Thomas Cole, Frederic Church, Albert Bierstadt, and Thomas Moran. These and others helped popularize a new and distinctly American cultural idea: the love of wild nature.

This love of wildness took on a deeply patriotic cast; many came to cherish it as something uniquely American. Artist Thomas Cole wrote in 1836: "Though American scenery is destitute of many of those circumstances that give value to the European, still it has features, and glorious ones, unknown to Europe ... the most distinctive, and perhaps the most impressive, characteristic of American scenery is its wildness."[5] The explorations of the continent and the reports back—the journals of Lewis and Clark, the paintings of George Catlin, and the reports of John Wesley Powell and early military expeditions, the images of pioneer landscape photographers such as William H. Jackson, to list just a few—fired the American imagination with the unique natural splendors of the young nation's wilderness.

In tracing the intellectual history of the very idea that the Grand Canyon is, in fact, grand, historian Stephen Pyne notes that early Spanish explorers saw nothing valuable or beautiful there, for they saw it through sixteenth-century eyes and for them, "the Renaissance was still aglow." They had "no evolved aesthetic or science for canyons," whereas

those who came three hundred years later followed a scientific revolution in natural history, a Romantic revolution in art and consciousness, and a wave of democratic political revolutions. For them ... the abruptness of Canyon scenery became a tenet of its appreciation, part of an aesthetic canon of Canyon mannerisms, and the abruptness of its rims, an illustration of the erosive mechanics that had sculpted the region.[6]

As Professor Pyne notes, the timing of the exploration of the wild American West "allowed the earth sciences to intersect the Romantic syndrome of the age," a process that "emerged piecemeal, as inherited ideas worked through new information."[7]

John Muir

In the final years of the nineteenth century, the great exponent of wild nature was John Muir, a figure of immense national influence through his writings and, in his later years, through his leadership of the Sierra Club, which he and other influential Californians founded in 1892. Though largely self-taught, Muir was a crucial transitional figure, linking threads of cultural and artistic thinking exemplified by Emerson and Thoreau with the scientific thinking rising in influence in his own time. His patient working-out of the glacial origins of Yosemite Valley went head-to-head against the theories of the most prominent trained geologists of the time, and Muir was ultimately proven right. In 2004 Muir's image and that of Yosemite Valley were chosen as the epitome of California's identity for its quarter coin in the series of state quarters.

Muir's prolific and evocative writings spoke to something deep in the psyche of the still-young nation with its vivid memories of the closing frontier. "Everybody needs beauty as well as bread," he wrote, "places to play in and pray in, where nature may heal and give strength to body and soul alike."[8] The theme of finding God in God's wilderness was ever present in his popular essays and best-selling books. "No

---

Walk away quietly in any direction and taste the freedom of the mountaineer. Camp out among the grass and gentians of glacier meadows, in craggy garden nooks full of Nature's darlings. Climb the mountains and get their good tidings. Nature's peace will flow into you as sunshine flows into trees. The winds will blow their own freshness into you, and the storms their energy, while cares will drop off like autumn leaves. As age comes on, one source of enjoyment after another is closed, but Nature's sources never fail. Like a generous host, she offers here brimming cups in endless variety, served in a grand hall, the sky its ceiling, the mountains its walls, decorated with glorious paintings and enlivened with bands of music ever playing.

—John Muir,
Our National Parks

synonym for God is so perfect as Beauty," he wrote, setting a theme that has resonated in wilderness literature and advocacy ever since.

> Whether as seen carving the lines of the mountains with glaciers, or gathering matter into stars, or planning the movements of water, or gardening—still all is Beauty! In God's wildness lies the hope of the world—the great fresh unblighted, unredeemed wilderness. The galling harness of civilization drops off, and wounds heal ere we are aware.[9]

In effect, the intellectual and cultural transformation in the eighteenth and nineteenth centuries put man on a course toward rediscovering that he is not the master of nature but a part of its fabric, dependent on it in myriad ways we still only superficially understand. It was a truth that ecological science would ultimately confirm. The transcendentalism of Emerson and Thoreau, the transcendence in Muir's writings, the insights of science—all were leading the same place, toward a recognition that we would do well to protect the remnants of America's wilderness before they were swallowed up by the maw of an older conception of "civilization" and "progress."

President Theodore Roosevelt and John Muir on the rim of Yosemite Valley, 1903. Muir, ever the propagandist for wilderness, hoped to "do some forest good in freely talking around the campfire." The two camped alone, reveling in the four inches of snow that fell on them overnight. Roosevelt returned to Washington, D.C., exulting that it was "the grandest day of my life!" (Photo © Bettmann/CORBIS)

Wilderness Vanishing

John Muir was a crucial transitional figure in another way, for he linked his nature-love to passionate advocacy for preserving national parks and forests. With his leadership, the Sierra Club became a convocation of wilderness-saving champions. His writings increasingly expressed deeply angry calls to action to save wilderness—calls for action by the federal government.

> Any fool can destroy trees. They cannot run away; and if they could, they would still be destroyed,—chased and hunted down as long as fun or a dollar could be got out of their bark hides, branching horns, or magnificent bole backbones. Few that fell trees plant them; nor would planting avail much towards getting back anything like the noble primeval forests. During a man's life only saplings can be grown, in the place of the old trees—tens of centuries old—that have been destroyed. It took more than three thousand years to make some of the trees in these Western woods,—trees that are still standing in perfect strength and beauty, waving and singing in the mighty forests of the Sierra. Through all the wonderful, eventful centuries since Christ's time—and long before that—God has cared for these trees, saved them from drought, disease, avalanches, and a thousand straining, leveling tempests and floods; but he cannot save them from fools,—only Uncle Sam can do that.[10]

We are part of the wilderness of the universe. That is our nature. Our noblest, happiest character develops with the influence of wildness. ... With the wilderness we are at home.

Some of us think we see this so clearly that for ourselves, for our children, our continuing posterity, and our fellow men we covet with a consuming intensity the fullness of the human development that keeps its contact with wildness. Out of the wilderness, we realize, has come the substance of our culture, and with a living wilderness—it is our faith—we shall have also a vibrant culture, an enduring civilization of healthful citizens who renew themselves when they are in contact with the earth.

—Howard Zahniser,
"Wilderness: How Much Can We Afford to Lose?"
Sierra Club Wilderness Conference, 1951

Muir was right to sound the alarm, for wilderness was fast disappearing. The drive to open and tame the last wild expanses of the once all-wilderness continent was at full momentum, hugely accelerated by the coming of the automobile. In 1925 Aldo Leopold wrote

> This country has been swinging the hammer of development so long and so hard that it has forgotten the anvil of wilderness which gave value and significance to its labors. The momentum of our blows is so unprecedented that the remaining remnant of wilderness will be pounded into road-dust long before we find out its values.[11]

Bob Marshall echoed Leopold, writing in 1937: "wilderness is melting away like some last snowbank on some south-facing mountainside during a hot afternoon in June. It is disappearing while most of those who care more for it than anything else in the world are trying desperately to rally and save it."[12] In his classic, *A Sand County Almanac*, Leopold wrote that in the rush to open every last bit of undeveloped land, the wild things wilderness lovers sought have eluded their grasp, and they hope "by some necromancy of laws, appropriations, regional plans, reorganization of departments, or other form of mass-wishing to make them stay put."[13]

The early history of wilderness preservation can be seen as an increasingly frantic effort to find the most effective "form of mass-wishing" to make wilderness "stay put," even as it was rapidly "melting away." It was the nexus of these two new ideas—that wilderness is not a luxury or mere playground but a fundamental human need and that by the 1920s and 1930s wilderness was vanishing "with appalling rapidity"—that energized the founding of the modern wilderness movement. Before exploring that phase, however, we need to understand why national parks, a hugely popular American innovation, ultimately did not prove to be an adequate vehicle for wilderness preservation.

National Parks and Wilderness

John Muir placed great hopes in the national park idea to protect wilderness. In the years after the establishment of the world's first national park, Yellowstone, in 1872, he was the greatest evangelist of national parks. His ceaseless advocacy for parks in his writings, together with the tourism promotion by the western railroads and local Muir-like advocates (such as Enos Mills and his advocacy for Rocky Mountain National Park), was the great driver of expansion of the number of national parks in the late nineteenth and early twentieth centuries.

National parks were the first concerted federal effort to preserve expanses of wild land for their natural values, although from the outset the greatest premium was placed on scenic spectacles and "wonders"—the geysers, the precipitous valleys and canyons, the towering glacier-clad mountain peaks.

However, many wilderness advocates soon soured on national parks as a means of really preserving wilderness. From the very start, the national parks mixed the goal of protecting nature with promoting more and more tourism by providing increasingly comfortable access and facilities as elaborate as those in any small town. In 1921 Leopold wrote that while it might be well and proper that "the Parks are being networked with roads and trails as rapidly as possible," because of tourist boosterism, the parks could not be the means of preserving true wilderness.[14] In 1925 the same idea was voiced in a leading scientific journal by Dr. Charles C. Adams:

> The older champions of our national parks, [such] as John Muir, were among the leaders in this country to see in a broad way the value of preserving wild areas, but in recent years there has been an intensive movement to get vast crowds into the national parks, and at such a rate that vast areas of the parks are without question being severely injured.[15]

This dissatisfaction was shared by park and wilderness advocates across the country. The frustration was expressed by a leader of the Federation of Western Outdoor Clubs in a letter to the National Park Service in 1939: "We have seen many of our favorite trails turned into roads and highways, our remote camp spots into auto camps and cabin grounds. ... There were beautiful mountain flowers where those cabins, auto camps and roads now lie. ... "[16] He added that wilderness advocates were no longer willing "to take a chance on the National Park set-up" to preserve wilderness.

Attitudes shared by many early leaders of the National Park Service only encouraged these apprehensions. They sought to have it both ways, promoting ever more park development yet protesting that the parks were, in fact, preserving wilderness values—even as the extent of that wilderness was being nibbled and sometimes gobbled away. From the outset they dug in their heels to resist drawing any actual line, even if only by administrative order, that would specify the wild parts of the parks they promised never to develop. Typical of their concept-only, no-firm-boundary-lines promises was a 1928 article in *The Saturday Evening Post* by National Park Service director Horace M. Albright:

> Thus was born the idea of the national parks, perpetual wildernesses, the last remnants of Nature's handiwork on this teeming earth. They are to be preserved forever in their natural state for the benefit and enjoyment of the people, to use the exact words of the act of Congress of 1872, creating the Yellowstone National Park. ... Of course the parks should remain wildernesses. It is true that they are the only primeval areas protected by law from the ravages of civilization. They must be saved as such.[17]

Wilderness advocates were not persuaded. Despite such promises to hold onto wilderness values in the parks, in 1937 Bob Marshall complained that with their enthusiasm for trail building, machinery, and large crowds, "the Park Service seems to have forgotten the primitive."[18]

The question conservation leaders increasingly asked in the 1920s and 1930s was how wilderness could be more deliberately and more permanently preserved on America's vast federal lands, whether it was within national parks, national forests, or other jurisdictions.

PART 2: THE BEGINNINGS OF A WILDERNESS PRESERVATION POLICY

Aldo Leopold

The essential leader of the wilderness preservation cause in the 1920s was Aldo Leopold. Celebrated today most of all for his exposition of ecological thinking and the conservation land ethic, Leopold shaped many of his ideas of ecological wholeness early in his career, conceiving and popularizing a practical program for wilderness preservation while serving as a district ranger with the U.S. Forest Service in the Southwest.

In the *Journal of Forestry* in 1921, Leopold praised the progress of national forest development but noted that it

> has already gone far enough to raise the question of whether the policy of development ... should continue to govern in absolutely every instance, or whether the principle of the highest use does not itself demand that representative portions of some forests be preserved as wilderness.[19]

Aldo Leopold on a 1924 canoe trip in what is now the Boundary Waters Canoe Area Wilderness of northern Minnesota. "The number of adventures awaiting us in this blessed country seems without end," he wrote in his journal. (Courtesy of the University of Wisconsin Madison Archives)

What was needed, Leopold urged, was "a definite national policy for the permanent establishment of wilderness recreation grounds."

In 1925 Leopold spelled out his wilderness ideas in more detail in the *Journal of Land and Public Utility Economics*. He urged that wilderness be seen as a natural resource in its own right, not solely as a collection of raw materials to be developed but rather as "a distinctive environment which may, if rightly used, yield certain social values." Saving wilderness should be, he urged, a part of a balanced pattern of land uses. Leopold expanded on a key idea, that wilderness has a fundamental patriotic value in American culture:

> There is little question that many of the attributes most distinctive of America and Americans are the impress of the wilderness and the life that accompanied it. If we have any such thing as an American culture (and I think we have), its distinguishing marks are a certain vigorous individualism ... a certain intellectual curiosity bent to practical ends, a lack of subservience to old social forms, and an intolerance of drones, all of which are the distinctive characteristics of successful pioneers.[20]

Leopold's concern was that if the rapidly vanishing remnants of wilderness were lost, then its influence in our national story and culture would soon fade, relegated to history books.

Are these things worth preserving? This is the vital question. I cannot give an unbiased answer. I can only picture the day that is almost upon us when canoe travel will consist in paddling in the noisy wake of a motor launch and portaging through the back yard of a summer cottage. When that day comes, canoe travel will be dead, and dead, too, will be a part of our Americanism. Joliet and LaSalle will be words in a book, Champlain will be a blue spot on a map, and canoes will be merely things of wood and canvas, with a connotation of white duck pants and bathing "beauties."

The day is almost upon us when a pack-train must wind its way up a graveled highway and turn out its bell-mare in the pasture of a summer hotel. When that day comes the pack-train will be dead, the diamond hitch will be merely rope, and Kit Carson and Jim Bridger will be names in a history lesson. Rendezvous will be French for "date," and Forty-Nine will be the number preceding fifty. And thenceforth the march of empire will be a matter of gasoline and four-wheel brakes.

—Aldo Leopold,
"Wilderness as a Form of Land Use,"
Journal of Land and Public Utility Economics, October 1925

Leopold did not limit his advocacy for wilderness to internal agency discussions and professional journals, but espoused wilderness preservation for a wider audience in popular magazines such as *Sunset* and *American Forests & Forest Life*.

The First Wilderness Area and the "L-20" Regulation

In June 1924 Leopold convinced the regional forester in the Southwest to establish the Gila Wilderness Area of some half-million acres. Though this was only an administrative order by a regional official and made no promise of permanence, it was nonetheless a landmark in American history: the first wilderness area officially designated as such by the federal government. Leopold's idea took hold, echoed by others within the Forest Service staff. Soon other wilderness areas were established by regional agency officials. But Leopold put this initial progress in perspective for readers of *Sunset* magazine in 1925: " ... let no man think that because a few foresters have tentatively formulated a wilderness policy, that the preservation of a system of wilderness remnants is assured in the National Forests."[21]

In his annual reports in 1926 and 1927, the chief of the Forest Service, William B. Greeley, said the wilderness idea had "merit and deserves careful study" and that he was making plans "to withhold these areas against unnecessary road building" and protect their wilderness character.[22] Greeley ordered an inventory of undeveloped national forest areas which might be suitable wilderness areas, and this work, undertaken by Leon F. Kneipp, led to Forest Service promulgation of the "L-20" regulation in 1929, the first nationwide policy for wilderness preservation. It adopted the new official term "primitive area" for these lands, applying it to the Gila and other areas already established by regional orders. The stated purpose of L-20 was "to maintain primitive conditions of transportation, subsistence, habitation, and environment to the fullest degree compatible with their highest public use with a view to conserving the values of such areas for purposes of public education and recreation."[23]

Notable though these steps were, they were only first steps, and hesitant at that. The primitive areas were no real answer to the goal of preserving wilderness in perpetuity. Nothing about the L-20 regulation actually prohibited any form of development or use, including road building and logging. Forest Service instructions sent to the field with the L-20 regulation specified that

> the establishment of a primitive area ordinarily will not operate to withdraw timber, forage or water resources from industrial use, since the utilization of such resources, if properly regulated, will not be incompatible with the purposes for which the area is designated.[24]

Life-or-death decisions for wilderness were left largely to the discretion of Forest Service field personnel, many of whom had no particular sympathy for the wilderness idea.

Bob Marshall

In the 1930s, as Aldo Leopold became engrossed in research, a new champion took up the leadership of the wilderness movement: the enormously energetic Robert (Bob) Marshall. Marshall was an extraordinary apostle, preaching the cause of wilderness and gathering acolytes wherever he went in his far-flung travels. He was independently wealthy and a trained resource professional with a forestry degree and a Ph.D. in plant physiology who had reveled in exploring the wilderness of New York's Adirondack Mountains in his youth. Like Leopold, he was a writer, publishing one of the seminal articles in wilderness history, "The Problem of the Wilderness," in the *Scientific Monthly* in 1930, as well as a best-selling book, *Arctic Village*, evoking the year he spent living in a small village amid Alaska's Brooks Range wilderness.

In his 1930 article Marshall argued that the fate of the wilderness would be decided within the next few years, posing a question of "balancing the total happiness which will be obtainable if the few undesecrated areas are perpetuated against that which will prevail if they are destroyed." He quoted some of Leopold's ideas, stressing that "as long as we prize individuality and competence it is imperative to provide the opportunity for complete self-sufficiency" beyond "the effete superstructure of urbanity," in "the harsh untrammeled expanses." Like Muir, he evoked the "dynamic beauty" of wilderness in contrast to a Beethoven symphony, "a landscape by Corot or a Gothic cathedral." Wilderness, he wrote, is in constant flux:

Bob Marshall on Mount Doonerak, Alaska, in 1939. Marshall wrote "no comfort, no security, no invention, no brilliant thought which the modern world had to offer could provide half the elation of twenty-four days in the little explored, uninhabited world of the arctic wilderness." Doonerak is now part of the wilderness within Gates of the Arctic National Park, part of the largest unbroken expanse of wilderness yet preserved by law on the planet. (Courtesy of the Bancroft Library, University of California, Berkeley)

A seed germinates, and a stunted seedling battles for decades against the dense shade of the virgin forest. Then some ancient tree blows down and the long-suppressed plant suddenly enters into the full vigor of delayed youth, grows rapidly from sapling to maturity, declines into the conky senility of many centuries, dropping millions of seeds to start a new forest upon the rotting debris of its own ancestors, and eventually topples over to admit the sunlight which ripens another woodland generation.[25]

Marshall urged that a study "should forthwith be undertaken to determine the probable wilderness needs of the country," saying it ought to be a radical calculation, in part because

the population which covets wilderness recreation is rapidly enlarging and because the higher standard of living which may be anticipated should give millions the economic power to gratify what is to-day merely a pathetic yearning. Once the estimate is formulated, immediate steps should be taken to establish enough tracts to insure every one who hungers for it a generous opportunity of enjoying wilderness isolation.[26]

In the mid-1930s Marshall became chief forester of the Office of Indian Affairs in the Department of the Interior, where he peppered Secretary Harold Ickes with memoranda promoting wilderness preservation. In 1934 he urged the secretary to order "a nationwide wilderness plan" to protect areas in the national forests, national parks, on lands administered by the Bureau of Land Management, and Indian lands.[27] A month later he sent Ickes a memorandum outlining a "Suggested Program for Preservation of Wilderness Areas" and, in the best New Deal spirit, urging creation of a Wilderness Planning Board to select the areas. To assure maximum protection, he advocated that the selected areas be designated by act of Congress, noting, "This would give them as close an approximation to permanence as could be realized in a world of shifting desires."[28]

The Wilderness Society

Even as he agitated for wilderness within the government, Marshall and his nationwide circle of wilderness friends knew that wilderness needed a strong push from outside forces. He had asserted in his 1930 article: "There is just one hope of repulsing the tyrannical ambition of civilization to conquer every niche on the whole earth. That hope is the organization of spirited people who will fight for the freedom of the wilderness."[29]

Fulfilling this need, Marshall was the central force around which The Wilderness Society was organized in 1935. With his independent means, he was also its main financial supporter. Over the next decades The Wilderness Society was the focal point for the growing national wilderness movement. Its governing council and staff were a rare brain trust, bringing together key thinkers of the wilderness movement. Just imagine the fine wilderness talk among the eight founders—Marshall, Aldo Leopold, Benton MacKaye, Robert Sterling Yard, Harvey Broome, Ernest C. Olberholtzer, Bernard Frank, and Harold C. Anderson.

The annual meetings of this extraordinary leadership group and their rich correspondence, together with their frequent interactions with agency officials, leading scientific and intellectual figures of the day, and leaders of other conservation groups—particularly during the biennial wilderness conferences the Sierra Club began holding in 1949—became the crucibles of wilderness thought and action in the 1940s, 1950s, and 1960s.

Forest Wilderness or National Parks?

By the 1930s the Forest Service was well embarked on its wilderness program using regulation L-20, while the National Park Service was pushing for establishment of new national parks. This set up a fierce bureaucratic rivalry between the Forest Service (in the Department of Agriculture) and the National Park Service (in the Department of the Interior). The rivalry was inflamed by Secretary of the Interior Harold Ickes's undisguised efforts to transfer the Forest Service to the Interior Department, a move bitterly resisted by the agency and its timber- and grazing-industry constituents. Blocked from his bigger ambition, Secretary Ickes countered with an aggressive campaign to transfer wild portions of the national forests piecemeal—notably the primitive areas—by converting them to national parks.

One Forest Service partisan, H. H. Chapman of the Yale School of Forestry, reported this rivalry in the *Journal of Forestry* in 1938 in specific wilderness terms:

> A terrific drive is on by the National Park Service, to capture and capitalize the sentiment back of the wilderness idea, and with this backing to secure as parks, the 11 million acres of wilderness or "primitive" areas already established within the National Forests. ... One of the most pressing arguments used is the assumed precarious status of any area set aside for a wilderness solely by executive orders of the [Chief] Forester or Secretary of Agriculture. ... For this reason I [feel] that wilderness areas should be given legal status by acts of congress, but that on the other hand ... they should remain as integral portions of the National Forests and not be transferred and take park status, to be subjected to the pressure for development which is desecrating so many of our most prized National Parks like the Yellowstone and Sequoia.[30]

Chapman neatly captured both sides of the argument. To the Forest Service, the best defense against National Park Service takeovers was to argue that in national parks wilderness would be "desecrated," to use Chapman's choice of words, by overdevelopment to encourage mass tourism—overdevelopment encouraged by the National Park Service. To park advocates, on the other hand, Forest Service primitive areas had no real security and could be abolished at the stroke of an unelected administrative officer's pen, whereas national parks had park protection by statutory law.

Wilderness advocates astutely took advantage of this rivalry, recognizing that fights over which agency could better protect wilderness would only encourage stronger wilderness policies by both. Consider the role of Bob Marshall. In 1934, when he worked in the Department of the Interior, he wrote to Secretary Ickes about "the prompt action required if we are to save the seven genuine wilderness areas in the National Forests which are in imminent danger of destruction. I would suggest that the U.S. Forest Service be asked for an immediate expression regarding the possibility of preserving these areas as wildernesses."[31] Marshall knew that the secretary would jump at the opportunity to promote some of these lands as national parks. Coming from Ickes, such a request about the status of national forest wild lands in the jurisdiction of a different and rival federal department could only inflame intense Forest Service alarm.

In 1937 Marshall transferred to a senior position in the office of the chief of the Forest Service, becoming responsible for the wilderness program. To Secretary Ickes, who disapproved of his move, he explained his deep concern that the national parks were being overdeveloped and the opportunity to protect real wilderness was elsewhere. At the Forest Service he peppered his superiors with advice about keeping national forest primitive areas from becoming national parks, suggesting a strategy of strengthening their protection as national forest wilderness. At the same time, in his private role as a Wilderness Society leader, Marshall endorsed the Olympic and Kings Canyon National Parks, testifying to Congress that the group had weighed the competing proposals of each agency and found parks would better protect wilderness values in these areas.[32]

The National Park Service won most of these battles. Congress established Olympic National Park in 1938 and Kings Canyon National Park in 1940, each embracing wild lands that had been national forest primitive areas. Each had been the subject of national controversy and intense congressional lobbying, with the "who-can-best-protect-wilderness" arguments dominating the debates. Secretary Ickes sought to deflect the "parks-will-be-overdeveloped" Forest Service defense by announcing his policy that the wilderness of the new parks would not be developed, while he lambasted the weakness of the Forest Service's wilderness efforts. One Ickes ally, the sponsor of the Kings Canyon park bill, attacked the inconstancy of Forest Service primitive area protection:

> [The chief of the Forest Service] told us when he testified that it is by his order a wilderness area now. Yes, gentlemen, he made it a wilderness area with a stroke of his pen. His successor, with another stroke of another pen

can undo all that he did. There is no security when your rights are based on regulations that may be made today and repealed tomorrow. ... If you make this a national park the maintenance of the wilderness characteristic ... will not be dependent upon the whim of any man but will be fixed by law.[33]

That last point was not exactly true, since the National Park Service had adamantly refused to make any lasting commitments about exactly how much of the new parks it would promise to preserve permanently as wilderness. So, to strengthen his case in the Kings Canyon debate, Secretary Ickes went a step further. He issued a press statement at the end of 1938:

Since 1935, the Olympic National Park and Isle Royale National Park have been established. Both will be maintained as wilderness areas. ... [There is a] need for a greater stability of policy than can be insured by administrative orders. Areas dedicated as wilderness national parks should be protected forever by provisions of law designed for that purpose. This in addition to the protection all national parks receive by law against commercial activities.

I shall welcome it if the Congress of the United States will define and set standards for wilderness national parks, as well as provide for wilderness areas to be proclaimed and similarly protected by law in other national parks.[34]

Secretary Ickes then had his staff draft a generic national park wilderness protection bill and promote it to key members of Congress. His bill was introduced by senior leaders of both the House and Senate in early February 1939—it was the first national wilderness legislation. The proposed legislation set the goal of preserving "perpetually for the benefit and inspiration of the people of the United States the primitive conditions existing within national parks and national monuments," authorizing the president to proclaim "wilderness areas when he determines that it would be to the public interest to do so."[35] Apart from its merits, the bill was an adroit legislative move typical of Ickes, a stratagem acting to counter the Forest Service line of attack about overdevelopment of wilderness in national parks.

The Forest Service sought to trump this tactic by writing Ickes to request that he also include national forests in the bill. Not surprisingly, as the debate over Kings Canyon swirled on in 1939, the Department of the Interior rebuffed the Forest Service request. Though a Senate hearing was held, Ickes's legislation ultimately languished, having served its strategic purpose during the final stages of the Kings Canyon national park fight, which he won.

"Wilderness" and "Wild Areas"—The "U" Regulations

Bob Marshall was well positioned as head of the lands division of the Forest Service to lobby internally for more protective, more permanent wilderness protection regulations. There was an evident political need for the Forest Service to offer a more credible promise of wilderness protection in the face of the intense threat of additional national park proposals. During the final months of the Kings Canyon national park fight, in mid-1939, Marshall reminded his superiors "I have been contending for over a year and half that if we didn't put our house in order there would be a good chance we would lose this area."[36]

Through these internal discussions, Marshall and allies such as Leon Kneipp made progress, if only slowly. Finally, in September 1939 the secretary of agriculture revoked the Forest Service L-20 primitive area regulation, replacing it with new departmental-level regulations "U-1" and "U-2." These created a new set of labels— "wilderness areas" for those 100,000 acres and larger and "wild areas" for those from 5,000 to 99,999 acres. Management directions, which were identical for both categories, were tightened to better protect wilderness values. Logging and road building were excluded. The permanence of the larger wilderness areas was enhanced to some degree by elevating their establishment and any subsequent boundary changes as a matter for higher-level decision by the secretary of agriculture. This had the effect of making any decision to reduce an area more visible and, therefore, perhaps less likely. The L-20 primitive areas altogether amounting to 14 million acres—were given the new, stronger protection temporarily, but each was to be reevaluated through a reclassification process, with public hearings and reconsideration of its boundaries, before it would be given the new wilderness or wild area titles. In practice, this reclassification process dragged on into the 1960s; of the original 14 million acres more than 5 million were still awaiting this reclassification when the Wilderness Act became law in late 1964.

Conservation groups welcomed the newly strengthened Forest Service policies, but over the next two decades they were increasingly disenchanted by the crawling pace of the promised reclassifications. More alarming was a pattern of decisions that realigned the boundaries, removing areas of timber from one primitive area after another for logging. In fact, between 1939 and 1964 the total acreage protected by the Forest Service for wilderness purposes barely grew—from 14.2 million acres to 14.6. And this 2 percent increase in net acres masked a trend of reduced biological diversity, as more rocky peaks and ridges were added to compensate for the removal of forests for logging.

Chapter 3

Wilderness: "There Ought to Be a Law"

How Permanent?

Bob Marshall died suddenly in November 1939; he was two months short of his thirty-ninth birthday. (He had suffered mysterious hospitalizations over the years; some thought it was heart trouble. The autopsy found myelogenous leukemia and coronary arteriosclerosis.)[1] His death shocked the small wilderness movement, which found itself without the charismatic leader whom Robert Sterling Yard had called "the most effective weapon of preservation in existence."[2]

With Marshall gone, the intellectual and strategic leadership for wilderness shifted even more from the federal agencies to citizen advocacy groups. The leaders of The Wilderness Society, Izaak Walton League, National Parks Association, National Wildlife Federation, the Sierra Club, and local leaders they worked with around the country were a small, close-knit fraternity. They increasingly saw the Forest Service and other federal land-management agencies as inherently too beholden to commercial constituencies and too easily swayed by shifting political pressures to be institutionally capable of really preserving wilderness in perpetuity.

Less than a year after Marshall's death, Harvey Broome, a Wilderness Society founder, wrote to his fellow Wilderness Society council member Olaus Murie, a prominent wildlife biologist and wilderness advocate. Broome explicitly linked the loss of Marshall's dependable leadership to this growing concern about the permanence of agency designations:

> After Bob Marshall died, I came to realize, and I guess it is the experience of all members of the [Wilderness Society] Council, how much I relied upon his immense knowledge. Since his death I have been wondering just how permanent and legally inviolable are the various wilderness areas in this country. What is to hinder some future Secretary from abrogating those [U-1 and U-2] Regulations? ... Do you think wildernesses would have more permanence if there were some new status, established by congressional enactment?[3]

In the national parks it increasingly seemed to be open season on wilderness, with conservationists rushing from one defensive battle to another. Commercial interests proposed logging in Olympic National Park. Federal dam-building agencies were planning new reservoirs seemingly everywhere. The National Park Service hierarchy, afire with enthusiasm for recreational development, was actively planning ever more tourism facilities. Notwithstanding Secretary Ickes's efforts, wilderness advocates remained deeply skeptical of agency promises to protect park wilderness.

After Marshall's death, the Forest Service added no significant amount to the 14.2 acres of primitive areas that had been protected at the end of 1939. And as the 1940s and 1950s unfolded, wilderness proponents' faith in the newly strengthened level of administrative protection for national forest wilderness collapsed in the face of contrary on-the-ground experience. Each time one of the old primitive areas was reclassified as wilderness, the boundaries were revised. On the basis of an exhaustive review for his dissertation, Dr. James Gilligan told the society of American Foresters in 1954 that

> there is a national trend of wilderness boundary modification which, since 1940, has eliminated over a half million acres of land from 33 different units. Most of these deletions have been for ... timber harvest, motorized recreation development, intercity road construction, or areas where mining or tourist facilities have already been established on private land. These deletions have largely been offset by the addition of high, rocky zones to each area where there is little possibility of development demands or timber harvest. It places a peculiar emphasis on maintaining a large national acreage figure for the wilderness system. ... [4]

Even though the gross acreage might remain the same or even be expanded a little, the quality and biological diversity of wilderness was whittled away. It was, as Leopold said, "a paper profit and a physical loss."[5]

Wilderness advocacy leaders saw both the Forest Service and the National Park Service weakening wilderness protection in case after case, siding with commercial interests to lop off a bit of wilderness here, another bit there, all too often encouraging development that eroded roadless wild lands. The "melting away" of remnants of wilderness Marshall had warned of was accelerating. Referring to the Forest Service policy, John Barnard of the Sierra Club observed on the eve of the Wilderness Bill campaign in 1956 that

Eventually master plans were prepared for national parks showing the ultimate limit of planned developments, but in the framework of this planning, wilderness seemed to be viewed mainly as the land left over in planning. Rather than being positively identified as a value in its own right, wilderness became the residuum in master planning.

—Michael McCloskey,
"The Wilderness Act of 1964: Its Background and Meaning,"
Oregon Law Review, June 1966

many people will be surprised that protection of America's [national forest] wilderness rests on so slim a base as these regulations. ... Theoretically, America's wilderness areas could be wiped out by the stroke of the Secretary of Agriculture's pen even though public sentiment did not favor it.[6]

"Not by Law"

To wilderness leaders, it was unacceptable to see wilderness policy lurch along in such an impermanent, haphazard way. Wilderness had to be saved, securely, once and for all—in perpetuity. A stronger means simply had to be found to save it, something inherently more permanent than the unfettered administrative discretion that both the National Park Service and the Forest Service so jealously guarded. Wilderness advocates wanted protection less prone to behind-the-scenes political alteration.

Wilderness leaders struggled with these concerns, sharing the attitude expressed by Izaak Walton League leader Kenneth Reid in 1939:

There is no assurance that any one of them [forest or park wilderness areas], or all of them, might not be abolished as they were created—by administrative decree. They exist by sufferance and administrative policy— not by law.[7]

So, what was the outlook for America's wilderness in 1940? And what was the posture of the wilderness movement? We can take a snapshot:

Outlook for Wilderness 1940

- There was no national policy to preserve wilderness—no recognition of wilderness itself as a resource of value, no commitment to protect wilderness across all the federal agencies that managed wild lands.
- There was no definition of wilderness—no agreed-upon practical standard applied by all agencies.
- There was no nationwide system of wilderness areas—a few agencies delineated a few wilderness areas; others ignored the wilderness values of their lands.
- There was no consistent management guidance for areas of wilderness.
- Wilderness preservation policy, such as it was, was still dominated by the federal agencies—who all too often bowed to powerful development or parochial political pressures.
- There was absolutely no commitment to preserve wilderness in perpetuity.
- And the wilderness movement, still very small, was stuck on the defensive, trying to hang on to remnants of wilderness.

The Idea of a Comprehensive Wilderness Program

Facing an utterly unacceptable outlook for the permanence of wilderness, through the 1940s activist leaders increasingly discussed finding a comprehensive, positive program for a stronger, better answer for wilderness preservation.

Aldo Leopold had been the first to promote—in 1925—a vision of a "definite national policy," advocating a nationwide system of wilderness in both the forests and parks. Marshall built on this—in 1934—with his vision for "a nationwide wilderness plan" involving lands in the national forests, national parks, the public domain lands administered by what is now the Bureau of Land Management, Indian reservations, and state and private lands. Today, this idea of a single national system of protected wilderness seems perfectly obvious to us, but it was not always so. In the early years of the twentieth century, as historian Donald Swain notes, Robert Sterling Yard, then president of the National Parks Association, "masterminded a kind of rhetorical strategy designed to build a feeling of unity toward the parks. By originating terms such as 'National Park System,' he helped reduce the chances of one unit in the 'system' being successfully attacked."[8] And in his 1934 memo to Secretary Ickes, Bob Marshall suggested the need for protecting wilderness areas by statutory law, a point echoed in Kenneth Reid's 1939 Izaak Walton League article.

Howard Zahniser

Robert Sterling Yard died in 1945 at age eighty-four, working to the last from his apartment bedroom as the president of The Wilderness Society and editor of its magazine. Into the leadership of the wilderness movement came Howard Zahniser, a man whose personality and skills turned out to be ideal for that juncture in wilderness advocacy. A charter member of the society, Zahniser was a professional editor with a scholarly turn of mind, recruited from a secure government job to step into Yard's role in Washington, D.C. He was teamed with Olaus Murie, who served as half-time director based in Moose, Wyoming. Harvey Broome later eulogized "this brilliant team" under whose leadership "the concept of wilderness became an imperative in American life."[9]

As Zahniser and Murie began their tenure, pressures to exploit the resources on federal lands were rapidly accelerating under the postwar economic boom. In 1945 Murie sized up the threat for wilderness:

> According to present plans, we must not remain satisfied with America as we have known it. The blueprints virtually postulate "lifting the face of Nature." The new dams proposed are so numerous they can not be conveniently listed. It is known that many of them would affect wildlife resources to an alarming degree. Many, with accompanying roads, would invade wilderness areas. Never before has there been a greater threat to what remains of primeval America.[10]

Facing this threat, Zahniser's vision—built on the yeasty discussions within The Wilderness Society's governing council—was, almost from the start of his work in 1945, the vision of wilderness preserved by law—a federal wilderness law.

Howard Zahniser, one of the least known yet most influential conservationists of the twentieth century, during a 1955 pack trip in the Selway-Bitterroot Wilderness, Idaho and Montana. The horses got loose and took off, except for this one who wanted to see what "Zahnie" was writing in his journal. (Courtesy of Ed Zahniser; James Marshall, photographer)

Imagining a Wilderness Law

As he began his job, Zahniser soon gained his initial experience with wild land legislation. A bill had been introduced in 1945 to protect a national system of trails modeled on the Appalachian Trail. In the words of the bill, these were to be "constructed, developed and maintained in a manner which will preserve as far as possible the wilderness values of the areas traversed."[11] In 1946 Wilderness Society founder and president Benton MacKaye (the father of the Appalachian Trail) broadened this legislative idea, asking the sponsor to introduce a revised concept for a nationwide system of "wilderness belts." Wilderness belts would be designated along wild rivers and crests of mountain ranges. MacKaye explained to Representative Daniel Hoch, the sponsor, that this melded the trail and wilderness ideas: "national *system* vs. regional, and *area* vs. line (or trail)."[12] MacKaye expressed the deeper political potential to Zahniser, whom he had asked to deliver his letter and the draft legislation. While the wilderness belts idea went no further, MacKaye's vision would be achieved two decades later in the Wilderness Act:

We have here the chance perhaps to launch a constructive national legislative campaign. ... It would make a definitive manoeuver to shift our ground from the defense to the offensive—from less negative to more positive action.[13]

Facing a growing tide of postwar developmental pressures, the governing council of The Wilderness Society met in 1947 at Rainy Lake in northern Minnesota. There they formally adopted Benton MacKaye's recommendation to focus the society's work on securing a nationwide system of wilderness areas more securely protected by statutory law. This was a carefully discussed departure, made with their eyes open to the challenge they faced in getting Congress to take such a proposal seriously. They well appreciated that they were making a fundamental shift, away from a process of decisions dominated by the federal agencies and toward an approach in which the politics inherent in the legislative process would shape wilderness decisions. They judged this the better hope for securing wilderness and gaining, as Bob Marshall had phrased it more than a decade earlier, "as close an approximation to permanence as could be realized in a world of shifting desires."[14]

To imagine a wilderness law, much less to enact it, required supreme boldness and foresight. In that period the political obstacles facing any such proposal were indeed daunting. David Brower, who became the first executive director of the Sierra Club soon after World War II and was Zahniser's close partner in this work, would later recall that many at the time thought it "was political madness ... to take on so many [likely] opponents at once."[15] Zahniser understood that enacting such a law would take unparalleled effort and support from many more voices than the few thousand members of organizations that gave wilderness issues top priority. There was no meeting of the minds among the broader community of national conservation groups about the need for such a law, nor was there agreement about what such a law would include. Obviously a more concerted effort and sophisticated strategy were needed before drafting a law that might win approval.

Someone must be able to visualize how society can deal comprehensively with its environment before the other prerequisites can acquire a practical relevance. This vision is less an act of individual inspiration than the slow and random accumulation of concepts and ideas from many sources that one day fall into place as a coherent and pervasive doctrine of social responsibility. To make this vision meaningful and to catalyze consensus is the function of leadership.

—Lynton K. Caldwell,
"Administrative Possibilities for Environmental Control,"
Future Environments of North America

As a preliminary step in his campaign, Zahniser felt it would be helpful to have an authoritative, neutral study to help bolster the case. He arranged for a congressman to request such a study from research experts of the Library of Congress to see "whether or not there is a need for a national wilderness policy; and, if so, what form it should take."[16] The library's Legislative Reference Service sent out a questionnaire, seeking opinions from many quarters; the completed study was published with impressive cachet as a congressional document in 1949. The author, economist C. Frank Keyser, concluded

> Opinion is general that a simple, clear definition [of wilderness] should be decided upon that is not too restrictive. ...
>
> A national-wilderness policy would probably require national legislation. The policy should have the same status as the national-parks policy. ...
>
> Permissible uses should be spelled out, as well as those which should be prohibited. ...
>
> Nearly all [of the agencies and organizations surveyed] agree that permanence of the wilderness nature of these areas must be provided for in any statement of national policy, most of them believing it should be to the same degree as that assuring the permanence of national parks.[17]

Building Consensus

In preparing The Wilderness Society's input for the Library of Congress study, Zahniser worked out in his own mind, more comprehensively than ever, the concepts for a practical and politically viable wilderness law and the case for such a proposal.[18] His lengthy memorandum is a masterful statement of the rationale for stronger wilderness preservation.

After discussions with many advisors, including friends in the federal agencies, Zahniser first laid out his case for a wilderness law in a speech, "How Much Wilderness Can We Afford to Lose?," delivered at the Sierra Club's Second Biennial Wilderness Conference banquet in 1951:

> Let's try to be done with a wilderness preservation program made up of a sequence of overlapping emergencies, threats and defense campaigns! Let's make a concerted effort for a positive program that will establish an enduring system of areas where we can be at peace and not forever feel that the wilderness is a battleground.[19]

"As soon as we have a clear consensus of conservationists we should most certainly press steadily for the maximum security possible; that is, congressional establishment of a national wilderness system backed by an informed public opinion," he said.[20] He outlined in eight points what such a bill should accomplish (and his points read essentially as a table of contents of the Wilderness Act signed into law thirteen years later). But it turned out that 1951 was not the right moment to launch the positive campaign he had been painstakingly preparing. For at that juncture along came a wilderness issue of surpassing political importance. Confronted by this new issue, Zahniser recognized that in campaigning to win it, he and his fellow conservation leaders could deliberately build the political leverage and the stronger coalition a realistic campaign for wilderness law would require.

Echo Park

In a plan that became public in 1950, the Bureau of Reclamation and western water interests proposed to build a dam and reservoir at a place called Echo Park on the Green River, a tributary of the Colorado. It was to be one element of a billion-dollar scheme to build ten federal dams across the Colorado Plateau.

Before they could launch the positive campaign for the Wilderness Bill, wilderness advocates were confronted with a proposed dam on the Green River, one mile downstream behind the left border of this photo from the period. The Echo Park dam would have drowned beautiful Echo Park, foreground, submerging the lower two-thirds of Steamboat Rock. (Courtesy of National Park Service, Historic Photographic Collection, Harper's Ferry Center)

The Echo Park fight arose because this dam was to be located in the undeveloped backcountry of Dinosaur National Monument. The coalition of development interests and politicians making up the powerful western water and development lobby saw no reason to deter them from putting one of their dams in this obscure unit of the National Park System. At that time Dinosaur National Monument was known—if known at all—for its dinosaur bones, not its remote wilderness river canyons. Yet where this proposed dam and this wild river collided, in a spectacular canyon setting known as Echo Park, fundamental conservation history was made.

The "Echo Park fight" was a black-and-white issue, a showdown between dam builders and wilderness advocates. From the outset Zahniser and his colleagues carefully defined the issue in fundamental terms—as a test of whether our nation would uphold earlier decisions to dedicate special lands for preservation or would abandon those decisions whenever some new development scheme came along. The conservation movement, then relatively tiny in size and resources, was seriously outgunned by the western water lobby and its congressional allies, backed by the power of the Eisenhower White House. Conservationists fought the proposed Echo Park dam as a matter of principle. Using Robert Sterling Yard's rhetorical stratagem, they argued that if a dam could be built in any one unit of the National Park System, even a unit as obscure as this, it would be a death blow to the integrity of the entire system, and thus to every national park and all wilderness, everywhere.

The Echo Park fight became, as one conservationist predicted at the outset, "a full-dress performance by all those ... who are protecting the West's recreational and wilderness values."[21] Richard Leonard, a member of both the Wilderness Society's governing council and the board of directors of the Sierra Club, pointed out: "[I]f we win on [the National Park System integrity] issue our battle may be won for the next 25 years, while if we win on other grounds (economic, for example) then we have not necessarily won a national park victory."[22] On the other side, Representative Wayne N. Aspinall, a prodevelopment Democrat who represented western Colorado, said, "If we let them knock out Echo Park dam, we'll hand them a tool they'll use for the next hundred years."[23] Both were correct.

The Echo Park campaign was to prove epochal in its impacts on the character and capabilities of America's nascent environmental movement. It became by far the most significant conservation lobbying campaign of the twentieth century. Later battles—and notably the campaign for the Wilderness Act—were won with strategic and tactical tools that Zahniser, the Sierra Club's David Brower, and their colleagues forged and perfected in the Echo Park campaign.

As historian Mark Harvey emphasizes in *A Symbol of Wilderness*, the definitive account of the Echo Park fight, this confrontation arose before the modern environmental movement had been armed with legal weapons to block the dam. It was a classic battle for public opinion: "As the controversy unfolded, Echo Park became a prime symbol of the American wilderness, at once magnificently beautiful as well as in danger of disappearing under development pressures."[24]

In a five-year, uphill, come-from-behind battle against all odds, conservationists won; the Echo Park dam was dropped from the package and a provision was added prohibiting dams in any national park unit.[25] In this victory, as historian Roderick Nash assessed it, the American wilderness movement had its finest hour to date. In the process, conservation leaders honed their own advocacy capabilities, becoming adept with newly sophisticated legislative strategy and lobbying tactics. They forged strong new alliances, welcoming new allies in a broadened coalition for wilderness. And they cultivated many new congressional allies, including Representative John P. Saylor, a conservative Republican from Pennsylvania who emerged as Congress's leading national park, wild river, and wilderness champion. Brower, Zahniser, and other conservation group leaders developed a new level of impact in the national media and mastered new techniques for grassroots mobilization of public opinion. A postwar generation of new, young wilderness advocates was tested and shaped in the Echo Park campaign, emerging with unprecedented energy and political savvy.

In the aftermath of their Echo Park victory, wilderness advocates were viewed with new respect in the halls of Congress. As important, Zahniser, Brower, and their colleagues, as well as wilderness supporters everywhere, gained an entirely new level of self-confidence in their own advocacy abilities on the national scene and in the scale of pubic support they could rally for wilderness.

To Zahniser, the Echo Park fight had been "the test whether any designation can long endure. We passed that test."[26] With vision and foresight, he had worked from the outset to script the Echo Park campaign as the vehicle for building the broader coalition and political momentum needed to launch the positive campaign for wilderness legislation he had been preparing for most of a decade.

Potent Conservationist Pressures: Members of Congress are impressed with the greater effectiveness of the conservationist forces. Passage of the Upper Colorado River Basin measure without the controversial Echo Park Dam is proof of the power of the new-style conservationist. Every legislator felt the sting of a campaign by citizens of every age and from every economic group. Nature lovers, bird watchers, recreationists, and old line Pinchot-Theodore Roosevelt conservationists have lined up solidly to protect the nation's park and recreational resources. Congressmen no longer scoff at their power.

—"What's Happening in Washington,"
Lawyer's Weekly Report,
April 9, 1956

The "Wilderness Bill"

As the Echo Park campaign was reaching its climax in 1955, Zahniser used speaking opportunities to renew his call for a wilderness law, laying out the justification for wilderness legislation—for what would come to be called the "Wilderness Bill" (capitalized even in newspaper reports). That March Zahniser spoke to the fourth biennial wilderness conference of the Sierra Club. Of this meeting, club leader Charlotte Mauk wrote immediately afterward, "It was obvious that the individuals and groups present were ready to say 'O.K.—we understand one another now and we have a pretty good idea of what we want. Let's go after it!'"[27]

In May 1955 Zahniser addressed the National Citizen's Planning Conference on Parks and Open Spaces in Washington, D.C. He prepared this speech—titled "The Need for Wilderness Areas"—with utmost care. It was the most important public address in the history of wilderness preservation. We need, Zahniser said,

> areas of the earth within which we stand without our mechanisms that make us immediate masters over our environment—areas of wild nature in which we sense ourselves to be, what in fact I believe we are, dependent members of an interdependent community of living creatures that together derive their existence from the sun. By very definition this wilderness is a need. The idea of wilderness as an area without man's influence is man's own concept. Its values are human values. Its preservation is a purpose that arises out of man's own sense of his fundamental needs.[28]

"We need congressional action," Zahniser concluded. Outlining the fundamental provisions needed in an effective national wilderness law, he told the conference, "we should move forward as steadily as we can toward this action."

The agreement by which conservationists won the Echo Park campaign was formally sealed in an exchange of letters between Zahniser, as the leader of the conservation coalition, and Representative Wayne Aspinall, leader of the House coalition working for the Colorado River Storage Act, in late January 1956. A week later Zahniser wrote out the first draft of the Wilderness Bill in longhand on his son Ed's grade-school pencil tablet. His draft was soon typed and shared with close confidants—George Marshall, Dave Brower, Charles Callison, Stewart Brandborg, and Mike Nadel.

The Wilderness Bill went through seventeen drafts in that busy spring of 1956, with Zahniser as midwife for every word: perfecting the technical details, tightening wording, and incorporating suggestions from his confidants. In its statement of congressional policy and definition of wilderness, he sought to make the text itself convey the deeper values of wilderness, so that the text of the bill became its own best advocacy vehicle. While struggling with the more-complex technical language, he wrote to George Marshall, Bob's younger brother who became a leader of both the

Portions of Howard Zahniser's first draft of the Wilderness Act, written on his son Ed's pencil tablet in the first week of February 1956, days after Zahniser led the triumphant resolution of the Echo Park controversy. Revised through seventeen drafts that spring, the bill was introduced in the Senate on June 7, 1956. (Author's collection)

Sierra Club and The Wilderness Society: "I am no bill drafter. If I had to do this again, I would much prefer to state all this in iambic rhyming couplets or even in the sequence of sonnets, than attempt to do this in bill language."[29] In fact, the poet in Zahniser made him the ideal draftsman. He crafted, in the earliest drafts, the masterful, evocative phrases of today's Wilderness Act.

Zahniser was carefully building the public case for a wilderness law, making use of a favorite device to educate conservationists across the country. In June 1955 he enlisted an ally from the Echo Park campaign, Senator Hubert H. Humphrey, a liberal Democratic senator from Minnesota, to place the text of his 1955 "The Need for Wilderness Areas" speech in the *Congressional Record*. Zahniser and his assistant, Mike Nadel, then arranged to have this version reprinted in booklet form. Tens of thousands of these were mailed by Senator Humphrey's office to the members of conservation organizations, inviting readers to send Humphrey their comments on Zahniser's proposal for a wilderness law. Those comments, in turn, were collated by Nadel and became the subject of a follow-up Senate speech by Humphrey in February 1956, even as Zahniser was deeply immersed in the drafting process. That speech, too, was reprinted and mailed far and wide. In this way, and in conservation periodicals, nationwide grassroots support for the Wilderness Bill was being carefully cultivated.

All along Zahniser and his closest colleagues were methodically gathering a new consensus of support among leaders of America's conservation organizations. Reflecting on that work, George Marshall told Zahniser that spring

> It is amazing what you have accomplished educationally, literarily, editorially, legislatively, with administrators, and organizationally. Perhaps most significant is the way you have helped and taught wilderness preservers how to work together.[30]

In June 1956 the Wilderness Bill was ready. It was introduced in the Senate by Senator Humphrey and in the House by Representative Saylor, with additional cosponsors of both parties. Again, Zahniser and Nadel arranged to have Humphrey and Saylor's introductory speeches from the *Congressional Record*, which incorporated the full text of the bill itself, reprinted as booklets, tens of thousands of which were mailed to conservation group members. This was a conscious tactic to place the complete text of the Wilderness Bill itself, along with explanations of its provisions, carefully drafted by Zahniser, in the hands of conservation-minded Americans in every congressional district. The brand-new bill thus gained backing by well-informed activists in all parts of the country. It was smart preparation, for the legislation faced long odds in Congress.

The Eight-Year Legislative Odyssey

It took eight years to pass the Wilderness Bill. It was vehemently attacked from the outset by an onslaught from the U.S. Chamber of Commerce; resource extraction interests such as the logging, mining, and cattle grazing lobbies; and members of Congress from both parties who were sympathetic with these groups. Any early passage was out of the question once the bill was opposed by leaders of the Forest Service and National Park Service and the Eisenhower administration. Both agencies worked behind the scenes to block the bill, while the Fish and Wildlife Service urged the administration to back it. Proclaiming their strong general interest in wilderness protection at the first congressional hearings, the Forest Service and Park Service nonetheless offered many technical objections. Responding to these arguments, Wilderness Bill sponsors worked with Zahniser to meet reasonable objections. As a result the bill went through many revisions through the rest of the 1950s but was not approved by congressional committees.

In resisting the Wilderness Bill, agency leaders' underlying motive was to preserve their administrative discretion. The chief of the Forest Service told Congress that the bill would "tend to hamper the free and effective application of administrative judgment which now determines, and should continue to determine, the use, or combination of uses, to which a particular national-forest area should be devoted."[31] They jealously defended, in other words, their freedom to change wilderness boundaries and protective provisions.

To draw the line against further development and to stand by that decision regardless of intense pressures requires a high degree of institutional commitment. Such commitment is inconsistent with realities of how large hierarchical public agencies operate, given their multiple and conflicting mandates, their powerful development constituencies, and their own ever shifting political leadership, let alone pressures from an unsympathetic White House. In a 1954 paper that had a strong influence on wilderness advocates, Dr. James Gilligan dissected these problems,

---

Those who understand the problem of wilderness preservation on federal lands are convinced that Congressional action is necessary to retain wilderness areas for future generations. It is improbable, however, that Congressional action or tighter administration to retain important wilderness regions can be effected only with the support of uncertain and divided wilderness proponents. The democracy of providing access to every parcel of our public lands may triumph, but will future generations appreciate this particular brand of wisdom?

—Dr. James P. Gilligan,
"The Contradiction of Wilderness Preservation in a Democracy,"
Proceedings, Society of American Foresters, Annual Meeting, October 24–27, 1954

pointing to "the improbability of preserving a system of large wilderness areas within the national forests."[32] Gilligan also emphasized "the distinct lack of enthusiasm among many Forest Service employees for fitting wilderness preservation into basic multiple-use philosophy. Against local pressures for economic or recreation development, it is claimed, the efforts for preservation cannot be justified by the relative few who use wilderness regions."[33]

Explaining the rationale for eliminating administrative discretion over wilderness designation and protections, one of the act's leading sponsors, Senator Richard Neuberger, Democrat of Oregon, stressed that

> this bill in no way reflects on the wonderful career services which now are in charge of wilderness areas and similar outdoor realms, but it actually seeks to safeguard these splendid men and women from undue political pressure, no matter what the source.[34]

A 1956 conservation group pamphlet written to rally grassroots support underscored the point:

> Our rare, irreplaceable samples of wilderness can be diminished at the will of the administrator, without the sanction of Congress. ... under the bill Congress would protect the wilderness interior as well as the boundaries of all dedicated wilderness. This would strengthen the hand of the good administrator, steady the hand of the weak one.[35]

John F. Kennedy endorsed the Wilderness Bill as part of his election campaign in 1960. As a result, the Wilderness Bill became the flagship of the activist conservation agenda of President Kennedy's New Frontier. Strong advocacy for the bill by his new administration, led by Secretary of the Interior Stewart Udall and Secretary of Agriculture Orville Freeman, muted the earlier agency opposition. This tipped the balance in the Democratic-controlled Congress.

The [Wilderness] Bill is of primary importance to westerners. The vanishing wilderness is yet a part of our western heritage. We westerners have known the wilds during our lifetimes, and we must see to it that our grandchildren are not denied the same rich experience during theirs. This is why the West needs a wilderness bill.

—Senator Frank Church,
Congressional Record,
September 5, 1961

Even so, the Wilderness Bill was subject to intense congressional debate and review for another four years. A leading opponent in the Senate later recalled, "Perhaps there is no other act that was scanned and perused and discussed as thoroughly as every sentence in the Wilderness Act."[36] As Michael McCloskey, who became Brower's assistant at the Sierra Club, summed it up: "Some sixty-five bills were introduced ... Eighteen hearings were held, six in Washington, D.C., and twelve in the field. Thousands of pages of transcript were compiled, and congressional mail ran as heavy as on any natural resource issue in modern times."[37] Opposition continued from many commodity industry groups and allied members of Congress. One supporter of the act, Senator Robert C. Byrd, Democrat of West Virginia, observed that

> The objectors seem to consider the chances of exploitation or further development of remaining areas of wilderness better under administrative determination of land policies than under an act of Congress. They like the status quo better than the prospects of congressional endorsement of wilderness preservation.
>
> Now I believe these opponents are right in concluding that wilderness will be more surely preserved with this legislation than without it.[38]

The Senate passed the bill by an overwhelming seventy-eight to eight vote in September 1961 with the leadership of Senators Clinton P. Anderson, a New Mexico Democrat, and Frank Church, an Idaho Democrat, joined by many colleagues, notably Senator Thomas Kuchel, a California Republican who was the Senate minority whip. However, the following August the House committee chaired by Representative Wayne Aspinall approved a completely rewritten substitute version unacceptable to conservationists and the White House. Representative Saylor called Aspinall's version a "perversion of the wilderness preservation legislation." It was, he said, "indeed a substitute bill. For the preservation of wilderness it substitutes protection for exploiters of our wilderness areas."[39] Advocates and their congressional champions calculated that any compromise between the strong Senate version and

---

One could hardly listen in on any of the hearings without realizing that the very opposition of the special interests in itself compellingly argues the need for the wilderness bill. They must know that the present protection of wilderness is conveniently weak—weak enough to allow commercial exploitation of these dedicated areas without too much trouble.

—David Brower,
Letter to the editor of the *New York Times*, December 19, 1958[40]

Aspinall's extreme substitute would emerge far too weak and come at the price of unrelated public land "giveaway" provisions in his version.

Wilderness advocates mobilized their grassroots and media supporters to urge that the Aspinall substitute be rejected in favor of the Senate-passed version, hoping to do this during an open debate on the floor of the House of Representatives. Angered by the affront of having the Speaker of the House, a Kennedy ally, refuse to let his version come to the House floor for a vote under a parliamentary procedure baring any amendments, Aspinall excoriated the "many pressures ... unleashed by the propagandists for the wilderness preservationists and their newspaper allies in some of our big cities."[41] He refused to negotiate, and that effectively killed the bill in the 87th Congress.

The Senate acted quickly in the next Congress, again passing the Wilderness Bill in April 1963, by a seventy-three to twelve vote.[42] Immediately after the Senate acted, Kennedy administration leaders and Zahniser began working behind the scenes to seek some accommodation with Chairman Aspinall. While encouraging Representative Saylor to carry the banner for his own bill, which was very similar to the strong version passed by the Senate, Zahniser conferred with the White House and, with Saylor's knowledge, encouraged Representative John D. Dingell, a prominent Democratic conservationist from Michigan, to initiate quiet conversations with Chairman Aspinall to explore possible accommodations. Just before he departed for his fateful trip to Dallas in November 1963, President Kennedy—a friend of Aspinall's from their time together as junior members of the House of

In the Rose Garden, September 3, 1964. President Lyndon B. Johnson has just signed the Wilderness Act and hands pens to Margaret (Mardy) Murie (left) and Alice Zahniser. Their husbands, Olaus and Howard, had died during the final year of the long lobbying campaign. Secretary of the Interior Stewart L. Udall leans over the president as members of Congress look on: Senator Frank Church behind Murie; Representative Wayne Aspinall behind Zahniser; Senator Clinton P. Anderson to the right of Aspinall and directly above the president; and Representative John P. Saylor, with glasses, standing close to Udall. (Courtesy of National Park Service, Historic Photographic Collection, Harper's Ferry Center; Abbie Rowe, photographer)

Representatives in the early 1950s—was on the telephone with the chairman to confirm progress for the bill.[43]

Ultimately, a new version of the Wilderness Bill was agreed upon, based on a bill Dingell introduced to embody the accommodations made with Aspinall and the White House. As Brookings Institution scholar James L. Sundquist put it, "the result was hardly a compromise—it was, rather, an acceptance of Aspinall's terms."[44] These changes focused on Aspinall's insistence that new mining claims be allowed in national forest wilderness areas for an additional period of time and also reflected a different mechanism by which new wilderness areas could be added to the system, shifting from a kind of automatic process drafted by Zahniser to requiring a new act of Congress to add any lands. This version passed the House of Representatives in July 1964 by a vote of 374 to 1.[45] It was a compromise bill—Aspinall voted for it— but the fundamental architecture as envisioned by Zahniser was intact. A few final refinements were worked out by a House-Senate conference committee, and the final version passed both the House and Senate unanimously in August.

President Lyndon B. Johnson signed the Wilderness Act on September 3, 1964. It had been forty years since the first wilderness area, the Gila Wilderness Area in New Mexico, was established administratively by the Forest Service with far weaker protection.

Howard Zahniser was not at the Rose Garden ceremony to savor this historic achievement he had catalyzed; he died in his sleep on May 5, 1964, a week to the day after he testified at the final congressional hearing on the bill. His widow, Alice, stood with Olaus Murie's widow, Mardy, at President Johnson's side as he signed the Wilderness Act into law.

Zahniser's death at fifty-eight did not come as a great surprise to his closest confidants, for he had lost the function of one lung, which complicated his already compromised heart condition. As a measure of the depth of Zahniser's relationship with members of Congress, Senator Anderson interrupted Senate debate that morning to announce his death. Representative Aspinall gave his own eulogy on the House floor (one of many), joined the family at the funeral home, and attended the funeral. In Dave Brower's eulogy to Zahniser, published in the *Sierra Club Bulletin*, he observed of Zahnie's long campaign for the Wilderness Act:

> It was political madness, some political scientists observed, to try to take on so many opponents at once. They simply didn't have the measure of Howard Zahniser's skill as a constant advocate. ... What made the most difference was one man's conscience, his tireless search for a way to put a national wilderness policy into law, his talking and writing and persuading, his living so that this Act might be born. The hardest times were those when good friends tired because the battle was so long. Urging these friends back into action was the most anxious part of Howard Zahniser's work. It succeeded, but it took his last energy.[46]

Zahniser's dream was to see Congress enact a Wilderness Bill that would stand the test of time, truly perpetuating, in the words of the law itself, "an enduring resource of wilderness" for "the permanent good of the whole people." He earnestly sought to build the broadest possible consensus around the bill, to give its future implementation the most auspicious start.

We too often fall into the habit of using militaristic metaphors to describe political "fights" and "battles." Zahniser, however, was a lobbyist of a different stripe, rare then or now. Olaus Murie described his method of working: "eager to find an opportunity, taking advantage of every opening, always with good judgment in crises, and an unusual tenacity in 'lost' issues."[47] Zahniser sought to see the virtues in opponents' arguments, striving to find a way, if he could, to meet their objections. During the Echo Park campaign, his colleague Fred Packard of the National Parks Association commended Zahniser on his "happy faculty of standing firmly by your guns, and yet being able to give those on the other side of the line credit for integrity and sincerity."[48] This son of an evangelical Free Methodist minister did not see opponents as enemies; he genuinely coveted coming to some understanding whereby their reasonable interests and his for wilderness preservation could be reconciled. "We don't force—we persuade," he said, "we try to meet objections."[49]

During the years when Chairman Aspinall was the roadblock, the two men had good measure of each other; Zahniser and his allies had bested Aspinall and his on the Echo Park issue. Aspinall had held out for a hard bargain on the Wilderness Bill in the House showdown of 1962. When Aspinall was hospitalized with a heart attack in 1963, Zahniser sent him a book and note at the hospital, later writing to him: "I shall not cease to hope that you and I can live and work together ... to help achieve what I am sure both of us will be happy to see accomplished."[50]

One might think this was just a toadying letter cynically intended to play up to Aspinall, but not from Zahniser, who considered Aspinall a potential friend of wilderness whom he could ultimately convince. Zahniser's evangelical approach is reflected in a private letter he wrote to Harvey Broome, the president of The Wilderness Society, while Aspinall was hospitalized:

> I do hope that [Aspinall] is all right and will remain on the scene, for I have faith that we are going to be able to work things out with him. If so, our consensus will be the broader and our prospects the better [for preserving wilderness].[51]

So, with the ink drying on President Johnson's signature on the Wilderness Act in 1964, how had the outlook for America's wilderness been changed? We can take another snapshot:

Outlook for Wilderness 1964

- The Wilderness Act established a clear, unambiguous national policy to preserve wilderness—recognizing wilderness itself as a resource of value.
- The Wilderness Act established a specific definition of wilderness—a practical standard ready to be applied to real areas.
- The Wilderness Act established the National Wilderness Preservation System—immediately designating the first 9,140,000 acres in statutorily protected wilderness areas.
- The Wilderness Act set out a single, consistent wilderness management directive—guidance applying to wilderness areas in the jurisdiction of all federal land-management agencies.
- The Wilderness Act mandated a wilderness review process for additional federal lands—agency studies, preliminary proposals, local public hearings, leading to presidential recommendations to Congress, all in a ten-year timetable.
- The Wilderness Act asserted the exclusive power of Congress to designate wilderness areas.
- The Wilderness Act also asserted the exclusive power of Congress to decide on any reversals of wilderness designations or alterations of wilderness boundaries—taking that power out of the hands of executive branch administrators.
- The Wilderness Act thus constituted the best mechanism to actually preserve wilderness in perpetuity.
- And the Wilderness Act shifted the wilderness movement from the defense to the offense—establishing a practical means to extend statutory wilderness protection to more federal lands.

Chapter 4

Putting the New Wilderness Act to Work

The Law Changed Everything

Had citizen-activists not doggedly lobbied the Wilderness Act into law, how much wilderness would now be officially protected behind real boundaries in America? The answer, it is fair to say, would be dramatically less than the almost 106 million acres that made up the National Wilderness Preservation System in mid-2004 (with additional areas pending congressional approval). It is a fair question to ask how much once-protected wilderness a later administration would have thrown open to development if doing so had remained within their executive powers, not requiring congressional decision. The 1964 act changed everything about wilderness preservation in the United States. It:

- Proclaimed a national policy to preserve "an enduring resource of wilderness" in perpetuity
- Designated the first 9.1 million acres of *statutory* wilderness areas as the down payment toward our growing National Wilderness Preservation System
- Mandated a ten-year review process, with the federal agencies to study and report to Congress on hundreds of potential wilderness areas, including:
  - All roadless portions of units of the National Park System
  - All roadless portions of units of the National Wildlife Refuge System
  - The 5 million acres remaining in thirty-four national forest primitive areas still not reclassified from their 1939 status
- Shifted the power to decide which areas to protect—and to determine the boundaries—from the agencies, instead requiring an act of Congress for any new wilderness designations or boundary alterations
- Proclaimed a common mandate for wilderness stewardship, applying to all designated wilderness areas on all types of federal lands. The core of the stewardship obligation is the congressional mandate to preserve the "wilderness character" of each area

Wilderness advocates understood that decisions made by Congress would be inherently political, but their experience showed that decisions made by the political appointees in the secretive upper echelons of federal agencies and departments (which are always subject to White House influence) are no less political, just less visible and less open to effective citizen influence.

Advocates also knew that the labyrinthine process of enacting a law affords disproportionate opportunities for its opponents to delay or kill a bill, and that compromise is inherent in legislative decisions. After all, these same advocates had utilized such tactics to block the Echo Park dam in the early 1950s and to block Chairman Aspinall's unacceptable substitute version of the Wilderness Bill in 1962. And they had contended with their opponents who had used similar tactics to hold up the Wilderness Bill during its eight-year odyssey.

Zahniser, Brower, and their colleagues saw in these immutable facts of legislative life the key to preserving wilderness in perpetuity. For once Congress has designated a wilderness area, those seeking to reverse the designation or make boundary deletions face a heavy procedural burden. Inertia is inherent in the legislative process, making it far easier in most cases to kill legislation than it is to pass it. If conservation groups are opposed to such a boundary-changing bill, they will have the advantage of that legislative inertia in working to block it.

The Wilderness Act initiated three fundamental changes, more apparent now than at the outset, each a great portent for the future of wilderness:

Because of the Wilderness Act, the wilderness advocacy movement decentralized for greater political impact.

Congress is a national body, but on any parochial subject its members pay great deference to the views of the senators and representatives from the state involved. For the wilderness movement to be effective in gaining new designations, this legislative reality would require building citizen leadership within the local state and congressional districts involved in each new wilderness proposal.

Because of the Wilderness Act, Congress does not defer to agency wilderness recommendations, but reaches its own decisions.

Agency leaders do have favored access and a powerful voice in congressional decision-making. Nonetheless, the Congress often designates wilderness areas the agency has recommended against and often expands agency-recommended boundaries when citizen groups make an effective case for the larger area.

Because of the Wilderness Act, stewardship of wilderness areas increasingly became a professional specialization and career focus.

Agency personnel draw on a wide range of knowledge to meet the unique challenges of preserving wilderness character. Training offered for agency personnel by the federal Arthur Carhart National Wilderness Training Center, which serves all four wilderness-administering federal agencies from headquarters in Missoula, Montana, promotes best practices in wilderness stewardship and encourages consistency across jurisdictions.

How has the Wilderness Act worked out in practice? By any measure, let alone in comparison with the dire outlook for wilderness preservation prior to the law, the progress of wilderness preservation has been impressive. Compare:

The Forty Years before the Wilderness Act

- From Aldo Leopold's success in gaining temporary designation of the Gila Wilderness in 1924 until the passage of the Wilderness Act four decades later, wilderness advocates and the Forest Service were primarily rearranging the boundaries of the same 14 million acres, the old national forest primitive areas administratively designated during the 1920s and 1930s. Wilderness lands in other agency jurisdictions had no formal protection at all.
- With the passage of the Wilderness Act, 9.1 million of those 14 million acres became "instant" wilderness areas, the first to receive statutory protection.

The Subsequent Four Decades

- By comparison, it has now been four decades since the enactment of the Wilderness Act. In this period our movement has secured additional statutory wilderness protection for almost 97 million acres, more than ten times as much as the original acreage protected by the law.
- In mid-2004, our National Wilderness Preservation System comprised 105,695,176 acres.

Data as of July 2004; www.wilderness.net

The Ten-Year Wilderness Review Process

The terms of the act mandated a process in which the Forest Service was to review the remaining 5 million acres of old 1930s primitive areas and the National Park Service and Fish and Wildlife Service were each to review the roadless lands in their jurisdictions for suitability for wilderness designation. These studies were to give the president information on which to base his own recommendations to Congress concerning each area. All of these presidential recommendations were to be sent to Congress by September 1974. In a time before public participation in such federal agency decision-making was common, each preliminary agency recommendation was to be subject to a public hearing in the state. Then a refined agency recommendation worked its way up the executive branch hierarchy to the White House.

In these provisions, the Wilderness Act mandated the most advanced set of environmental study and public participation requirements in any law up to that time. (The now-familiar government-wide environmental impact review procedures of the National Environmental Policy Act did not come until 1970.)

As it turned out, the ten-year wilderness review process was an invaluable stimulus for the wilderness movement for three reasons:

Correcting Misinterpretations of the Act

Thanks to the Wilderness Act, agency wilderness recommendations were no longer the final word. Higher officials in the departments and the White House could improve them, and they often did so when advocates met with them to lay out the case for more expansive boundaries proposed by local citizens. The president had an independent role to make his own recommendations, not necessarily those submitted to him by the agencies. Then, when each presidential recommendation reached Capitol Hill, the congressional review and amendment process created yet another kind of "court of appeals"; wilderness advocates could spotlight misinterpretations of the Wilderness Act and other erroneous rationales offered to justify inadequate proposals and boundaries.

This was a time of intense controversy over agency misinterpretations of the new law—misinterpretations labeled as such by Congress with the leadership of legislators who had themselves written and championed the Wilderness Act. These fundamental errors included the "purity theory" that Forest Service officials offered to justify excluding millions of acres of wild lands they claimed were not wild enough to qualify. The National Park Service proposed to exclude "threshold zones," great swaths of wild lands between the edge of park roads and developments and their preferred, smaller wilderness boundaries. Exclusions of such lands were seen by conservation groups not as thresholds to the wilderness beyond but as beachheads for possible future development. Exclusion of this kind would leave zones in which agencies would still exercise the discretion to develop, a discretion the Wilderness Act was all about removing.

In the years following the Wilderness Act, as Congress deliberated on new wilderness area recommendations, it worked to correct these agency attempts to exclude lands from wilderness protection and to straightjacket the Wilderness Act by misinterpretation. To its credit, the Fish and Wildlife Service, which had been more favorable to the Wilderness Bill from the outset, largely avoided these problems of misinterpretation as it implemented the wilderness review process for the national wildlife refuges.

Establishing Fundamental Precedents

As the earliest areas were added to the new wilderness system in the late 1960s and early 1970s, congressional champions of the Wilderness Act—including Senators Anderson, Church, and Lee Metcalf, Democrat of Montana, and Representatives Saylor and Morris (Mo) Udall, Democrat of Arizona—not only corrected agency misinterpretations, they also paid close attention to assure that fundamental precedents were established. For example:

- It was important to establish that Congress would not simply rubber-stamp presidential boundary recommendations nor exclusively consider recommendations from the executive branch. Representative Saylor won an amendment to the bill designating the first new wilderness area—the San Rafael Wilderness in California, designated in 1968—by adding a critical 2,200 acres. In 1972 Congress designated the Scapegoat Wilderness in Montana, the first of many wilderness areas that originated as a citizen proposal, not from an agency or presidential recommendation.

- When it designated the first Department of the Interior wilderness area in 1968, the 3,700-acre Great Swamp Wilderness in a national wildlife refuge in New Jersey, the House of Representatives chose to close a two-lane township road that was all that divided the proposed area into two units in the agency proposal. It required three years after the president signed the law designating the wilderness for the road to be closed, the right-of-way acquired, and the roadbed removed, along with utility lines and poles. Today, few users of the delightful woodland trail may know that it was once a well-used road.

Great Change in Wilderness Movement

The ten-year review of hundreds of potential new wilderness areas posed a mammoth task, compressed in time and locally focused, with required public hearings in each state coming up in rapid succession. This confronted the still-small wilderness movement with a profound challenge. As the Wilderness Act became law, The Wilderness Society's annual budget was just $200,000; the Sierra Club membership was still small and concentrated in California. Neither had a broad network of regional offices, and there were few specialized statewide wilderness advocacy organizations.

The wilderness movement had no real choice but to change. The ten-year wilderness review process necessitated that citizens do their own field studies to help them prepare their alternative boundary proposals when they found the preliminary agency proposals inadequate. Local people had to be rallied to do these studies, to attend and speak up at each public hearing, and to then follow through, pressing for congressional action. This was more than a small movement centralized in offices in Washington, D.C., and San Francisco could accomplish. The upshot was a deliberate effort to refocus the wilderness movement. It was this change that created the far-flung, highly decentralized wilderness movement of today, with many hundreds of state and local wilderness organizations independently focused on areas they know best and supported by national conservation organizations in advocating their own citizen proposals—not the other way around. In retrospect, this was a profound transition, for it ushered in an era of unprecedented growth in organized grassroots support for wilderness in every part of the country.

Stewart Brandborg

With Howard Zahniser's death, another gifted leader emerged to lead wilderness advocates in the transition needed to effectively implement the Wilderness Act. Stewart M. Brandborg, a Montanan trained in wildlife biology, was a Washington,

D.C., conservation staffer for the National Wildlife Federation when he was invited to join the governing council of The Wilderness Society in 1956. "Brandy" was among the first colleagues with whom Howard Zahniser shared his early drafts of the Wilderness Bill, and he was involved in lobbying from the outset of the eight-year campaign for the law. In 1959 he moved to the society staff as assistant executive director and, upon Zahniser's death, became executive director just as the Wilderness Act was becoming law.

Because Brandy had worked side-by-side with Zahniser in the lobbying drive for the Wilderness Bill, he was well aware of the biggest defeat conservationists felt they had to swallow to get the act passed by Congress. This was the insistence by conservative western members of Congress, led by Chairman Aspinall, that each proposed new wilderness area would have to be added to the system by affirmative congressional approval—by passing a new act of Congress.[1]

Zahniser's original bill, as it had twice passed the Senate, proposed an automatic wilderness addition mechanism by which federal agencies would make their wilderness recommendations for each area, with specific boundaries. When recommended for wilderness designation by the president (with any changes he might choose to make), each recommendation would automatically become the new wilderness area, with those recommended boundaries—unless the presidential proposal was overruled by a "veto" resolution passed by Congress.

In opposing this "legislative veto" concept, Wilderness Bill opponents focused the fight on this question of "affirmative action," as they referred to it. They felt Zahniser's automatic process abused the sole power of the Congress as stated in Article IV section 3 of the Constitution "to dispose of and make all needful Rules and Regulations respecting the Territory or other Property belonging to the United States." Senator Henry M. Jackson, a Washington state Democrat and a supporter of the bill, summarized the two sides of this debate in a letter to a constituent:

> The real fight is between conservationists' desire for a mechanism which will force Congress to act to keep an area out of the wilderness system, and the effort of the opponents to require Congressional action before an area gets in. The proponents want to prevent wilderness proposals from dying because Congress fails to get around to them. The opponents want to capitalize on delays and oversights to keep areas out. It is a struggle for tactical advantage.[2]

The "Great Liberating Force"

In 1961 and again in early 1963, the Senate decisively voted down the affirmative-action alternative requiring a new act of Congress to add any land to the wilderness system. However, wielding his absolute power as committee chairman to block the legislation, Aspinall made acceptance of the affirmative-action approach part of his price for allowing the bill to even reach a vote by the full House. By late 1963 the conservationists and their congressional champions had bowed to Chairman Aspinall's absolute insistence on removing Zahniser's automatic mechanism, which had twice been approved by the Senate.

Zahniser, Brandborg, and their colleagues soon recognized the silver lining to their widely perceived "defeat." They came to perceive the great potential benefit in the requirement that new wilderness areas be designated only by act of Congress. Looking back several years later, Brandborg reported to The Wilderness Society's governing council in 1968

[Aspinall's] "blocking effort," as we saw it at the time, has turned out to be a great liberating force in the conservation movement. By closing off the channel of accomplishing completion of the Wilderness System substantially on an executive level, where heads of [conservation] organizations would normally consult and advise [senior agency officials] on behalf of their members, the Wilderness Law, as it was passed, has opened the way for a far more effective conservation movement, in which people in local areas must be involved in a series of drives for preservation of the wilderness they know.[3]

> Development of local and regional leadership will afford an unequaled opportunity to gain public interest and involvement in the wilderness preservation effort. It would be difficult to conceive of a finer opportunity for carrying out an aggressive educational campaign for the wilderness purpose.
>
> —Stewart Brandborg,
> Acting Executive Director's Report,
> The Wilderness Society, 1964

Grassroots Organizing and Citizen Wilderness Proposals

With the enactment of the Wilderness Act, the Sierra Club, led by Brower and his assistant, Mike McCloskey (who succeeded Brower as executive director in 1968), and The Wilderness Society, led by Brandborg and his new assistant executive director, Rupert Cutler, geared up to deal with the hundreds of agency wilderness studies and public hearings required by the Wilderness Act. They had no real choice but to build decentralized capacity to meet this challenge. Grassroots organizing became the new focus, not just for the immediate public hearings but for the follow-through to build constituent support that would encourage local congressional delegations to back citizen proposals.

The press of the agency studies and hearings directly contributed to the rapid nationwide expansion of the Sierra Club's chapter and group structure, for it gave club members in most states a close-to-home organizing task—something I experienced as a chapter volunteer leader in Michigan at the time. Both the club and The Wilderness Society hired grassroots organizers and opened new regional offices as rapidly as their finances allowed. This was a burst of growth well before the first Earth Day in 1970 (an event that many mistakenly believe marked the beginning of the environmental movement).

This turn to intensive grassroots organizing and the encouragement of grassroots groups as the lead advocates for their own citizen wilderness proposals was key to the growing political impact of the wilderness movement through the later decades of the twentieth century. Working for their local wilderness proposals gave local activists a hands-on education in practical legislative politics and its crucial dependence on effective grassroots organizing. Proposals with solid constituent support were added to the wilderness system, while those without generally languished. These practical lessons were increasingly applied to the broadening agenda of the environmental movement.

I first learned these lessons of grassroots politics (and became involved with the club and the society) in Michigan, when by happy chance I was able to attend the 1967 Wilderness Act public hearing on the preliminary National Park Service proposal for wilderness in Isle Royale National Park, flying to the northern Michigan location in a plane with leaders of both groups. I am a confirmed fan of the Park Service and the parks as reservoirs of wilderness. It was a summer job as a park ranger that led me to change course in the middle of my college years to study park issues and forestry at the University of Michigan. Having spent a glorious summer week on Isle Royale in 1966 as part of my class work, it was obvious to me that there was no justification for excluding large portions of this virtually all-wilderness national park from Wilderness Act protection, as the agency proposed. It took a decade to complete, but we ultimately succeeded in convincing Congress to designate 98.7 percent of the land area of Isle Royale National Park as statutory wilderness.

At the outset of my career with wilderness advocacy groups, I was tutored about the vital importance of grassroots organizing to protect wilderness. Organizing trips to places as diverse as Dunsmuir, California; Roseburg, Oregon; and across North Dakota in the years just after the first Earth Day in 1970 confirmed everything I had learned from studying Howard Zahniser's approach and from three gifted mentors, Rupert Cutler and Stewart Brandborg at The Wilderness Society and Mike McCloskey at the Sierra Club.

The grassroots organizing process involved identifying conservation-minded local people, connecting them with each other, helping nurture or establish local wilderness organizations, and focusing them on forthcoming agency wilderness studies in their area. Training sessions equipped these volunteer activists to conduct field studies and work out their own citizen wilderness proposals, often as alternatives to inadequate agency recommendations. In that period, gifted wilderness organizers, both staff and volunteers working with The Wilderness Society and the Sierra Club, as well as some local organizations, built the modern wilderness movement from the ground up.

As the next step, in a program innovated by Stewart Brandborg, The Wilderness Society invited grassroots leaders to Washington, D.C., for "Washington Wilderness Seminars," providing intensive training in how Congress works—and how citizens can influence its decisions. Participants in one of the first of these, in 1969, got particularly good real-world training when the group was enlisted to help in the final lobbying for a key wilderness-expanding amendment to legislation designating the Desolation Wilderness Area in California near Lake Tahoe.

If, after the local public hearings, an agency persisted with an inadequate wilderness proposal, as they often did, we would bring local citizen-leaders to Washington and go over the heads of the agency, setting up meetings with their bosses in the upper levels of the Interior or Agriculture Departments. We were often able to get agency wilderness proposals improved and expanded when citizens were persuasive in this kind of intervention at higher levels. But if not, we could also go to the very top of the executive branch, taking our case to sympathetic officials in the White House. Through the Johnson, Nixon, and Ford administrations, this often resulted in improved wilderness recommendations before the president sent them to Congress.

The ultimate goal, of course, was to be sure Congress designated what conservationists judged to be the best, most expansive possible wilderness areas. Even as each wilderness recommendation was moving up the hierarchy in the executive branch and toward Capitol Hill, we worked with grassroots leaders and their own elected representatives, so that often the bills introduced by local members of Congress were for the more expansive citizen proposals rather than the inadequate agency proposals. Beginning in 1968, hundreds of volunteer grassroots citizen-leaders traveled to Washington for training sessions, to lobby their congressional delegations, and to testify for their own wilderness proposals. It is a process continuing to this day.

This quantum leap in volunteer-focused grassroots organizing and training was the "great liberating force" Brandborg had recognized and of which he was the foremost evangelist—not an exaggeration of the passion he devoted to developing a larger, more diverse, and better-trained grassroots wilderness movement across the country.

Congress Expands Agency Wilderness Proposals

Thanks to Chairman Aspinall's "victory" in requiring that each addition to the wilderness system be enacted by an act of Congress, Congress became the "court of appeals" where citizen groups could challenge inadequate agency wilderness proposals. Senator Church, who had led the Senate fight against the affirmative-action approach insisted upon by Chairman Aspinall, acknowledged as much in 1972, telling one of Aspinall's Senate allies

I recall I was not persuaded at that time that this was the necessary way of handling new additions to the wilderness [system] but from what I have seen since, I think you were quite right.

I am glad the act contains this provision because it enables us now to exercise an oversight that we otherwise might well have relinquished to executive discretion alone.[4]

Through the democratic processes by which it considers legislation, Congress has proven receptive to citizen wilderness proposals if they are effectively presented by credible local people backed by sustained grassroots support. Thanks to Aspinall's requirement, as we worked with grassroots activists on the earliest wilderness proposals in the late 1960s and 1970s, more often than not Congress expanded wilderness areas beyond the boundaries proposed by the agencies and president.

After the Wilderness Act became law, the first area Congress added to the National Wilderness Preservation System was the San Rafael Wilderness near Santa Barbara, California, one of the old L-20 primitive areas from the 1930s. Here is how Chairman Aspinall, who still controlled the House committee, reacted to what happened during the evolution of this first new wilderness designation:

This primitive area was 74,990 acres. I repeat: 74,990 acres. After the first [Forest Service] hearing, this was raised to 110,403 acres. Then ... it was brought to Congress with 142,918 acres ... [Representative John Saylor] succeeded in getting 2,200 more acres [added by amendment in Aspinall's committee], which makes a total of 145,118 acres ... May I say to my colleagues, if this is going to be the trend in our determination of whether or not primitive areas are to become wilderness areas, and if we are going to increase them by 100 percent, then my opinion is that creation of new wilderness areas in the future are going to be very few and very far between.[5]

Chairman Aspinall did not grasp how his and other Wilderness Bill opponents' success in insisting that Congress would have to pass a new law to designate additional wilderness areas would turn out to empower grassroots wilderness advocates. In fact, the San Rafael Wilderness, which Aspinall lamented was being established by the 1968 act at 143,000 acres, has been enlarged twice by Congress since its initial designation. In 2004 the San Rafael Wilderness comprised 197,000 acres—contrasted to the original 75,000-acre 1930s primitive area—and a further citizen-proposed addition of 64,500 acres was pending in Congress.

Eastern Wilderness and the "Purity Theory"

The Wilderness Act designated four areas in the eastern United States as "instant" units of the wilderness system, all on national forest land. One was the large Boundary Waters Canoe Area in Minnesota. The others were much smaller areas the Forest Service itself had administratively designated as wild areas through the U regulations. Like many portions of national forest wilderness and wild areas in the West, none of these three wild areas were pristine lands, but nature was healing the wounds of earlier logging and other human impacts.

For example, the Shining Rock Wild Area in North Carolina showed fading evidence of extensive railroad logging and a huge logging slash fire that occurred between 1906 and 1926, before it became national forest land.[6] In establishing the Shining Rock Wild Area, the chief of the Forest Service had obviously determined that notwithstanding this history of past human impacts, it was an exemplary candidate for a wild area, which the agency described as areas "ranging in size from 5,000 to 100,000 acres in which the *primitive environment* is protected and preserved."[7]

In fact, the evidence of human impact could be quite recent, even current. The formal proposal that resulted in Forest Service administrative designation of the Shining Rock Wild Area in May 1964 (just as Congress was finalizing the Wilderness Act and therefore closely scrutinizing such agency designations) pointed out that

> In determining the best and most logical boundaries for the Wild Area, it was necessary to include a portion of the drainage of Ugly Creek covered by a timber sale contract which expires December 20, 1963. About 500 MBF [thousand board feet] are left to be cut and the operation will be completed this year. The skid trails and log landings will be revegetated and otherwise treated as necessary to hasten natural recovery and prevent vehicular access.[8]

As it began to pass laws adding areas to the wilderness system, Congress took the same practical approach, designating wilderness areas embracing once-logged and roaded lands. It is clear from the legislative history of the Wilderness Act itself that Congress intended such formerly abused lands to be within the law's pragmatic designation criteria. Indeed, many areas mandated for study under the ten-year wilderness review had a history of past land uses and abuses, including Shenandoah and Great Smoky Mountains National Parks and the Great Swamp, Seney, and Moosehorn National Wildlife Refuges.

By contrast, once the act became law, the Forest Service chose to ignore all this and adopt their own new interpretation of the act. By the early 1970s the agency was feeling seriously threatened by growing grassroots pressures as activists bypassed the Forest Service to ask Congress to protect "de facto wilderness"—areas in all regions of the country that had never been administratively protected as "primitive areas." Agency leaders felt that congressional expansion of the wilderness system might get out of hand. As a means to help preclude this, they adopted the idea that no areas with evidence of any past human impacts could or should qualify as wilderness under the Wilderness Act. This posture came to be known as the "purity theory." Forest Service leaders choose to bring this issue to a head in 1971, proclaiming that no national forest lands in the eastern half of the United States could even be considered for wilderness designation.

In contesting citizen proposals to designate new wilderness areas in the eastern half of the country, the agency tried to convince Congress to adopt their purity interpretation—conservationists said misinterpretation—of the Wilderness Act's definition of wilderness. In fact, the real issue was larger than just the East, for if Congress accepted the purity theory, it would give the Forest Service a powerful means of minimizing the expanse of wilderness boundaries in national forests in the West, where lower-elevation valleys and accessible plateaus typically had a history of some past human impact.

The Eastern Wilderness Areas Act

As the vehicle to gather congressional support for their purity idea, the Forest Service drafted new legislation to create a separate and competing system of "wild areas east." Among other things, their plan would have allowed cutting trees to "improve" wildlife habitat and recreation. If Congress were to enact this legislation, by implication they would be approving the Forest Service purity interpretation of the Wilderness Act for application in the West as well as the East.

Associate Chief John McGuire announced this wild areas east idea to an auditorium full of wilderness advocates at the 1971 Sierra Club wilderness conference in Washington, D.C. Despite immediate opposition from The Wilderness Society and Friends of the Earth, the Forest Service continued to promote the concept, leading to introduction of the bill they drafted in Congress. The bill—"to establish a system of wild areas within the land of the national forest system"—was described by its lead sponsor as necessary because the eastern areas "do not meet the strict criteria of the Wilderness Act."[9] This bill actually passed the Senate in 1972 (having been brought to the Senate floor with no prior notice and no debate) and there was, for a time, a split on this issue among conservationists (a few organizations had actually participated in the behind-the-scenes bill drafting discussions with the Forest Service). Led by The Wilderness Society, wilderness advocates responded with their own bill, the proposed Eastern Wilderness Areas Act. This bill packaged together numerous citizen proposals for de facto wilderness areas on national forests in the East, Midwest, and South.

To wilderness advocates and their congressional allies, there were two fundamental issues at stake in the showdown between the Forest Service's wild areas east legislation and wilderness advocates' Eastern Wilderness Areas Act:

Refuting the "Purity Theory"

The whole debate turned on the purity theory. Introducing the Eastern Wilderness Areas Act in 1973, Representative Saylor told the House, "Mr. Speaker, I am the author of the Wilderness Act in this House. I know very well what it says and what it intended, and I know how it was intended to be applied in a practical program."[10] He challenged those "who tell us [the act] is too narrow, too rigid, and too pure in its qualifying standards" to allow any formerly abused lands or lands with present abuse that can be restored with time.[11]

> I fought too long and too hard, and too many good people in this House and across this land fought with me, to see the Wilderness Act denied application ... by this kind of obtuse or hostile misinterpretation or misconstruction of the public law and the intent of the Congress of the United States.[12]

Introducing the companion bill in the Senate, Senator Henry M. Jackson, a veteran of the Wilderness Bill debates and chairman of the committee dealing with wilderness, told the Senate

> There is ... also a further purpose behind this bill. As it will be considered by the Committee on Interior and Insular Affairs, we will focus on a most serious question of interpretation involving the integrity of the Wilderness Act and our wilderness preservation policy. A serious and fundamental misinterpretation of the Wilderness Act has recently gained some credence, thus creating a real danger to the objective of securing a truly national wilderness preservation system. It is my hope to correct this false so-called "purity theory" which threatens the strength and broad application of the Wilderness Act.[13]

Like Representative Saylor, Senator Jackson, together with his lead cosponsor, Senator James L. Buckley, a conservative Republican from New York, and other members of the committee took this issue very seriously. Like wilderness advocates, they were concerned about the Forest Service effort to bifurcate what Congress has intended to be a single *national* wilderness system. As Senator Buckley explained

> If we are to have a national system of wilderness areas, as the drafters of the Wilderness Act obviously intended, less than pristine standards would be necessary for practical application. As a basis for public policy I believe it would be a mistake to assume that the Wilderness Act can have no application to once-disturbed areas.[14]

During hearings on the issue, Senator Frank Church, a leader in enacting the

Wilderness Act, zeroed in on the central point, telling the chief of the Forest Service

> If we [adopt the Forest Service theory] we will be saying, in effect, that you can't include a comparable area in the West in the wilderness system. That is the precise effect of your approach, because you will have redefined section 2(c) of the Wilderness Act.[15]

The subcommittee chairman, Senator Floyd Haskell, a Colorado Democrat, added

> I think the cat is now out of the bag. I couldn't understand ... how you could possibly interpret your definition of wilderness the way you do. What I gather now is that you are afraid all of the area that qualifies under the [Wilderness Act] definition will be designated as wilderness areas. I couldn't agree more with Senator Church that it is up to Congress to decide which of those areas will be included. You seem to fear that just because an area meets the definition it will be included.[16]

Senator Church concluded the dialogue, telling the chief to "bring your proposals up here, Congress is the final judge of what goes into wilderness and what does not."[17]

Seeking to Avoid Confusion of Congressional Committee Jurisdictions

By Forest Service design, their wild areas east legislation found its way to the congressional agriculture committees, not the interior committees that had jurisdiction over the Wilderness Act. The House and Senate "ag" committees were familiar precincts to the Forest Service and Department of Agriculture lobbyists but less friendly territory to wilderness advocates. Chairman Jackson and his interior committee colleagues were concerned about protecting their committee prerogatives, seeing the interest of the agriculture committee as an unfriendly effort to poach on their committee jurisdiction. Neither wilderness advocates nor these senators had forgotten that they had had to rally to defeat a surprise last-minute procedural

---

The wild areas [east] system would serve the Forest Service cause well. It sets two Senate committees to fighting over specific wilderness proposals; it confirms the Forest Service "purity" argument; fragments the wilderness movement into regional factions with less influence, instead of a unified national movement ... and it gives the Forest Service new hope for stopping the citizens' wilderness proposals in every western state, from California to Montana.

—George Alderson,
"Eastern Wilderness Crisis," *Environmental Quality*, March 1973

motion agriculture committee leaders had proposed during Senate debate, as a move to block the Wilderness Bill.

After prolonged controversy, both on Capitol Hill and within their own ranks, conservationists united and defeated the wild areas east legislation, instead passing the Jackson-Buckley-Saylor bill. The Eastern Wilderness Areas Act, which was signed by President Gerald Ford on January 3, 1975, designated fifteen new wilderness areas in eastern national forests pursuant to the 1964 Wilderness Act and required wilderness studies for seventeen other de facto areas.

As Forest Service historian Dennis Roth recounts, the new law "has erroneously been called the 'Eastern Wilderness Act'; however, it has no title." This is not a trivial matter; using this erroneous title implies that some special, separate law was needed to protect eastern wilderness areas. In fact, the correct name is the Eastern Wilderness *Areas* Act. This title was mistakenly left out of the bill itself in a clerical error at the eleventh hour as it passed Congress in the hectic final two days of the session. The title "Eastern Wilderness Areas Act" was in the bill as it passed the Senate and as approved by the House committee. As a result of the last-minute clerical error—back in the days when "cut and paste" meant literally that—the law President Ford signed begins with section two. There simply is no section one where the title should have been.

With this law, Congress definitively rejected the purity theory. Nonetheless, the idea resurfaces from time to time, fastened onto by wilderness opponents as a pretext for challenging new areas.

Lest there be any doubt, the designation of formerly abused lands as part of wilderness areas is consonant with the founding philosophy of the wilderness movement. The founder of the Forest Service wilderness program, Aldo Leopold, wrote in *A Sand County Almanac*, "in any practical [wilderness] program the unit areas to be preserved must vary greatly in size and in degree of wildness."[18] Congress has consistently followed this pattern, both before and after the eastern wilderness debate in the early 1970s. As part of the text of the 1984 Vermont Wilderness Act, for example, Congress declared that

> The Wilderness Act establishes that an area is qualified and suitable for designation as wilderness which (i) though man's works may have been present in the past, has been or may be so restored by natural influences as to generally appear to have been affected primarily by the forces of nature, with the imprint of man's work substantially unnoticeable, and (ii) may, upon designation as wilderness, contain certain preexisting, nonconforming uses, improvements, structures, or installations, and Congress has reaffirmed these established policies in the designation of additional areas since enactment of the Wilderness Act, exercising its sole authority to determine the suitability of such areas for designation as wilderness.[19]

As will be detailed in Chapter 8, the acceptance of past land uses and abuses most emphatically does not constitute a precedent that would allow the introduction of new conflicting land uses or abuses within a wilderness area once it is established.

Notwithstanding this episode, the Forest Service has a proud history in wilderness preservation and has taken more initiatives than any other agency to give roadless lands it administers consideration for possible recommendation to Congress.

National Park Service Resistance to Wilderness Designation

The political leadership of the National Park Service was never enthusiastic about the Wilderness Act. As its required wilderness studies and proposals appeared during the act's ten-year review period, the Park Service adopted a pattern of excluding a great deal of wild land on the basis of their own criteria. In one example, Senator Philip A. Hart, a Michigan Democrat, pointed out that the agency's wilderness proposal for Isle Royale National Park

> would exclude ... more than 6,000 acres. ... Yet, all but a very minor portion of this area is entirely undeveloped, totally wild and natural. There are no provisions in the approved park master plan for any expansion of development into these areas. There is, in short, absolutely no reason for not assuring these key lands the full protection of the Wilderness Act.[20]

At the same hearing, Senator Frank Church lectured the National Park Service for

> setting the boundaries of its proposed wilderness units back from the edge of roads, developed areas and the park boundaries by "buffer" and "threshold" zones of varying widths. There is no requirement for that in the Wilderness Act. No other agency draws wilderness boundaries in this way, which has the effect of excluding the critical edge of wilderness from full statutory protection. ... In the absence of good and substantial reasons to the contrary—and I [mean] specific, case-by-case reasons—the boundaries of wilderness areas in national parks should embrace all wild land.[21]

The upshot was congressional rejection of this kind of wilderness-minimizing policy.

Over the past four decades there has been a lack of sustained effort by the National Park Service to advocate wilderness protection for park lands. A number of studies required by the Wilderness Act have never reached Congress. Indeed, for some parks the studies required by law have never been initiated. Many presidential recommendations for wilderness within National Park System units have been sent to Congress, but many have languished for decades—the presidential wilderness proposals for Yellowstone, Glacier, Rocky Mountain, and Big Bend National Parks, to name just a few. In many cases, reluctance about the proposed wilderness designation among a particular state's congressional delegation could have been overcome by spirited advocacy from the agency, adding to the voices of wilderness advocates.

As a result of congressional establishment of huge new National Park System units in Alaska in 1980, with immediate designation of wilderness within many of them, the National Park Service now administers more statutory wilderness than any other agency. It could overcome its reluctant past to be America's premier wilderness agency. As a one-time park ranger, I believe it should.

Wilderness in National Wildlife Refuges

As naturalist Lois Crisler wrote, "wilderness without its animals is mere scenery."[22] Expanding the thought in recounting her experience with wolves and caribou in the Brooks Range in the 1950s, she described "a great living whole, with its proper inhabitants going about their business."[23] The essential role of the Wilderness Act in the National Wildlife Refuge System was expressed by Representative John Dingell, a key player in enactment of the Wilderness Act and the leading congressional champion for the refuges.

> Mr. Speaker, we should all recognize that many forms of wildlife found on the national wildlife refuges absolutely require a wilderness condition in order to survive. ,,, Placement of these refuges in the national wilderness preservation system will assure their continued survival of the wildlife and the continuance of their contributions to the enhancement of the environment.[24]

For just this reason, the Fish and Wildlife Service welcomed the Wilderness Act. As with the parks, however, many refuge wilderness proposals—particularly for large refuges in Alaska—still languish. Here, too, hostility by a state's congressional delegation—particularly in Alaska—presently stands in the way. As I was writing this book, I attended an agency "summit" about the future of the refuges and was shocked that under the thumb of the George W. Bush administration, the word "wilderness" hardly appeared in hundreds of pages of draft papers prepared for the summit participants to discuss and finalize. Conservationists look for the day when the Fish and Wildlife Service will be freed from such political handcuffs and again able to express the importance of wilderness for wildlife.

Chapter 5

Expanding the Scope of Wilderness Preservation

The "60 Million-Acre Myth"

Since 1964 Congress has year after year chosen to expand the National Wilderness Preservation System. Some evidently view this as a threat. Four decades after the Wilderness Act became law, some who were not involved allege that in the early 1960s, members of Congress and citizen advocates of the Wilderness Act envisioned a wilderness system limited to about 60 million acres.

Those making this argument imply that protecting any additional wilderness areas somehow contravenes some limit contemplated in the original congressional vision or by its citizen-advocates. In an April 2003 letter, a number of conservative western Republican senators expressed it this way:

> The legislative history makes clear that when debating the Wilderness Act of 1964, Congress and the advocates for a National Wilderness Preservation System envisioned a system of approximately 60 million acres.[1]

These senators went on to lament, "unfortunately, Congress failed to legislate any limits and to date Congress has established more than 107 million acres of wilderness."[2]

In apparent eagerness to discourage or shut down further congressional designation of wilderness areas, some special interests and overly parochial politicians have an obvious goal: to leave as much federal land as possible available for their own development ambitions. They argue that a 60 million-acre ceiling was implied in the legislative history of the Wilderness Act. This assertion is untrue. Nothing in the Wilderness Act or its legislative history states or implies any congressional expectation about the ultimate size of the National Wilderness Preservation System. The sponsors, congressional committees, and citizen advocates most centrally involved had no intent to set such a limit.

The history of the Wilderness Act, the statements of its congressional champions and citizen-advocates, and the more than 100 additional wilderness designation laws Congress has enacted since 1964 show

- The members of Congress who enacted the 1964 law did not presume to dictate what the extent of the wilderness system might be in the future.
- The framers of the Wilderness Act designed a process by which our National Wilderness Preservation System can, over time, meet the desires of the American

people as expressed through their elected representatives.

- The extent of America's wilderness system is determined by the cumulative democratic decisions of Congress in enacting laws designating additional areas.

In fact, Congress could not impose such a limit even if it wished to; the present Congress cannot bind the decisions of a future Congress. Year after year, the Congress then in session is in full control of how large the wilderness system becomes.

Of course, during testimony and congressional debates on the Wilderness Act, estimates of the extent of the *initial* National Wilderness Preservation System were discussed. But these estimates referred only to those specific federal lands directly addressed by the terms of that act. For example:

- The Senate committee report estimated the acreage directly dealt with in the Wilderness Act itself—its immediate designations and the areas it required to be studied—as slightly less than 60 million acres.
- An estimate offered by Senator Anderson, the lead Senate sponsor in 1961, suggested that the review of those lands specifically addressed in the act might lead to a wilderness system, at that stage, of 35 to 45 million acres.
- The chief House sponsor of the act, Representative Saylor, and the leading citizen advocate, Howard Zahniser, both used tables of areas addressed in the bill, totaling 62 million acres.

In each case these estimates explicitly covered *only* lands dealt with by the direct terms of the 1964 Wilderness Act, as distinct from whatever other federal lands Congress might choose to add in the future. As part of their advocacy for the Wilderness Bill in those final stages, the lead sponsors, Senator Anderson and Representative Saylor, emphasized the limited reach of the bill. Representative Saylor explained that he did so to answer exaggerated claims about the impacts of the Wilderness Bill:

> The bill does not lock up 1 billion acres, 776 million acres, and all of the national forests. While such allegations may be absurd to some, a number of people have from time to time mentioned each of these items as being a consequence of the present legislation.[3]

Since the Wilderness Act became law four decades ago, Congress had chosen to go beyond the limited set of federal lands it designated or required to be reviewed in the initial law:

- With overwhelming bipartisan support, Congress has chosen to designate millions of acres as wilderness on categories of federal land that were not addressed in the 1964 act, such as the roadless portions of national forests beyond the few old 1930s primitive areas addressed in the act.
- Similarly, in 1976 Congress chose to extend the reach of the Wilderness Act to the largest category of federal lands, the public domain administered by the Bureau of Land Management. These tens of millions of acres were not mentioned in the 1964 act.

- Congress chose, in 1980, to designate more than 55 million acres of lands in Alaska as wilderness. Most of these lands, too, were not addressed in the 1964 act.
- Congress continues to designate additions to the wilderness system, responding to the wishes of local people as reflected in the advocacy of their own congressional delegations and with bipartisan support.

The history of these congressional decisions to expand the scope of the wilderness system beyond the lands dealt with expressly in the original Wilderness Act illuminates the politics and issues of wilderness preservation in the twenty-first century. It is important to review the three major categories—national forest roadless areas, Bureau of Land Management lands, and the special case of Alaska.

PART I: NATIONAL FOREST "ROADLESS AREAS"

Of the original 14 million acres of the old 1930s-era national forest primitive areas, at the time the Wilderness Act became law there were still about 5 million acres the agency had not completed reclassifying under its 1939 regulations. Additionally, the Forest Service had not extended any sort of administrative protection to most other national forest lands with strong wilderness qualities. Indeed, the net acreage of administratively protected wilderness grew by just 2 percent in that twenty-five-year period from 1940 until the enactment of the Wilderness Act.

These 5 million acres of old primitive areas were the only part of the 192 million acres of national forests that the Wilderness Act required to be studied. Yet, four decades later, Congress has designated nearly 28 million acres of national forest lands beyond the original 9 million acres it designated in the Wilderness Act itself.

Having been necessarily consumed with getting the Wilderness Act passed, conservationists nonetheless knew that other national forest lands had very high wilderness values and remained vulnerable to development pressures. These other suitable lands came to be called the de facto wilderness and, later, roadless areas. Howard Zahniser discussed this category of potential wilderness areas in his 1962 testimony on the Wilderness Bill and Dave Brower offered his favorite definition of de facto wilderness in a 1962 speech: "They are simply 'wilderness areas which have been set aside by God but which have not yet been created by the Forest Service.'"[4]

Given the political terrain in the late 1950s and early 1960s, conservationists and the sponsors could not hope to pass even the basic wilderness law if they were too expansive in the lands they listed for mandatory wilderness study. As Congressman John Saylor said in a 1963 House speech

There are other national forest areas that are in fact wilderness but have never been so classified for protection as such. Nothing in this bill would prevent the Secretary of Agriculture from considering such areas for

Historic Progress of Wilderness Protection			
Agency	At end of administrative establishment of primitive areas November 1939	As the Wilderness Act became law September 1964	As of July 2004
U.S. Forest Service (adding categories below)	14,235,414	14,617,461	35,041,353
Wilderness areas (administrative before 9/64; statutory after)	0	9,139,721	34,867,591
Primitive areas for study (interim protection as of 9/64)	14,235,414	5,477,740	173,762
National Park Service	0	0	43,616,250
U.S. Fish and Wildlife Service	0	0	20,699,108
Bureau of Land Management	0	0	6,512,227
Total Acres Protected (wilderness and primitive areas)	14,235,414	14,617,461	105,868,938
Number of states with areas	13	13	44
Boundary changes can be made	By administrative decision	Only by act of Congress	Only by act of Congress

NOTES AND DATA SOURCES

Primitive area acreage from U.S. Forest Service. These received statutory interim protection in the Wilderness Act, so are counted here as statutorily protected after 9/64. Only the Arizona portion of the Blue Range primitive area remains under this interim protection. July 2004 data from www.wilderness.net except Blue Range primitive area, from Web site of Apache-Sitgreaves National Forest. *Millions of acres of lands in all four agency jurisdictions are pending in presidential recommendations not yet acted upon by Congress or are in official wilderness study areas, and these categories are not counted in this table.*

preservation. Each area, however, will have to be the subject of further legislation in the future [to be designated as wilderness].[5]

As Saylor implied, once the Wilderness Act was in place, it was not surprising that local people around the country recognized it was a tool they could use to secure statutory protection for areas of this de facto wilderness.

Stewart Brandborg of The Wilderness Society predicted in 1964, even before the Wilderness Act became law, that "the main recourse of citizens who seek protection of these [de facto] areas lies in their appeals to members of Congress."[6] In the late 1960s, grassroots groups in Montana, Alabama, Oregon, West Virginia, and other states began developing their own citizen proposals for areas of de facto national forest wilderness, often spurred on by imminent Forest Service plans to open these wild lands to roads and logging. Some areas were in states with well-developed wilderness support and sympathetic congressional delegations, so soon bills were introduced in Congress. Citizens understood that they could go around the Forest Service and ask Congress to designate their citizen-proposed areas as wilderness.

Across the country groups of citizens are working skillfully … preparing inventories of potential wilderness areas in various … jurisdictions. … working in task forces exhibiting impressive professional talents, they are delineating outstanding de facto wilderness opportunities, refining proposed boundaries, and drawing up detailed maps and supportive documentation.

—Representative John P. Saylor,
Congressional Record,
October 14, 1970

Scapegoat: The First Citizen-Initiated Wilderness Area

The first citizen-initiated de facto wilderness bill introduced in Congress was the Scapegoat wilderness proposal in Montana, spurred by Forest Service plans to road and log an area directly south of the Bob Marshall Wilderness. Montana had a strong grassroots wilderness movement led by the Montana Wilderness Association, Montana Wildlife Federation, and organizations of wilderness guides and outfitters. And the Montana congressional delegation was sympathetic, including Democrats Senator Lee Metcalf, an original sponsor of the Wilderness Act, and Senator Mike Mansfield, the majority leader of the Senate.

The local citizens, led by a small-town hardware dealer named Cecil Garland and working with The Wilderness Society's Clif Merritt, conducted their own field studies and organized a strong base of grassroots support, which gave their citizen wilderness proposal excellent credibility. It was introduced as legislation in 1965 and, over dogged opposition from the Forest Service, the quarter-million-acre Scapegoat Wilderness was enacted in 1972.

Many in the Forest Service had a tough time adjusting to the new political realities of wilderness designation. In frustration, the regional forester in Montana asked rhetorically,

> Why should a sporting goods and hardware dealer in Lincoln, Montana, designate the boundaries for the 240,000-acre [Scapegoat Wilderness] ... ? If lines are to be drawn, we should be drawing them.[7]

But that, of course, was just the point of the Wilderness Act—the decision-making and the boundary drawing were taken over by higher authority, the Congress of the United States.

NEPA, Roadless Areas, and "RARE"

In the years after the Wilderness Act became law, Representative Saylor led or supported many efforts, such as the Scapegoat Wilderness legislation, to designate additional wilderness areas on federal lands not addressed in the 1964 law. In 1970 he sponsored a bill to designate eleven new wilderness areas totaling more than 900,000 acres on national forests in eight states. None of these lands were "primitive areas" required to be studied by the 1964 law; all were citizen proposals. Representative Saylor explained to the House

> It is widely recognized ... that other areas beyond those specified for study by the parent [wilderness] act may warrant and, for wisest stewardship, require similar legal designation. I refer to areas—particularly in our national forests—which, although not yet protected as wilderness by law,

nevertheless exhibit natural values, wildness, and solitude of great national significance. We should not neglect nor delay the identification and proper conservation of these areas—which our constituents call de facto wilderness—simply because unlike others being reviewed, they may not have been administratively protected prior to passage of the Wilderness Act.

It is clear that the Wilderness Act, although it does not require that these areas of de facto wilderness be reviewed, makes possible their placement in the national wilderness preservation system. And we would surely fail in our duty to conserve them wisely for their highest values if we did not use the procedures and strong preservation policies in the Wilderness Act to protect them.[8]

The Forest Service became increasingly alarmed by such citizen initiatives and the strong support they were gaining in Congress. And, as this was unfolding, President Richard M. Nixon signed the National Environmental Policy Act (NEPA) in January 1970. NEPA created new, court-enforceable procedures requiring federal agencies to consider alternatives to their major decisions, including the decision to open still-wild roadless lands for development. This and the political desire to get ahead of and regain "control" over the burgeoning de facto wilderness issue led the Forest Service to initiate the 1971–1973 Roadless Area Review and Evaluation (RARE), a process to inventory all national roadless areas and select some for more intensive wilderness study, modeled on the study process of the Wilderness Act.

Though the Forest Service applied their purity theory criteria during RARE and thus deliberately overlooked virtually all roadless areas in the East, South, and Midwest, the agency inventoried 56 million acres of roadless land. In his 1973 decision, the secretary of agriculture selected 12.3 million acres from this total as areas for further wilderness study. Not surprisingly, conservationists' greater concern was with the 44 million acres the Forest Service did not select for study but hoped to be able to develop. Conservation groups were determined that these de facto wilderness lands should have more thorough assessment and fairer consideration before they could be freed for development.

Ultimately, in settling a Sierra Club lawsuit applying the still-new requirements of NEPA, the Forest Service agreed to evaluate the wilderness potential of each roadless area in an environmental impact statement (EIS) before development could proceed. Both the RARE process and the ensuing efforts by the agency to use the then-rudimentary national forest planning processes to meet this EIS requirement were deeply flawed, spurring greater citizen dissatisfaction and fueling new demands for Congress to step in to protect more of these de facto wilderness areas. A congressional committee concluded that the RARE process

arbitrarily fragmented large roadless tracts into smaller units thus lowering possible points given to an area on the Forest Service's rating system for "solitude"; deducted rating points for areas containing commercial timber reflecting a Forest Service policy that designated Wilderness Areas should not have significant volumes of commercial timber—a policy which is definitely not contained in the Wilderness Act; and adopted a "purity" definition and concept of wilderness so stringent as to preclude most "de facto" wilderness from further wilderness study.

The latter concept of wilderness, the so-called "purity" issue, has involved extensive debate. Testimony ... repeated allegations that the Forest Service has been unduly restrictive in setting wilderness evaluation criteria which relied solely on the most stringent possible interpretation [of the Wilderness Act]. For example, instead of recommending further wilderness studies in areas where "the imprint of man's work [is] substantially unnoticeable" ... the Forest Service's 1972 RARE review generally disfavored areas where any trace of man's activities was present.[9]

The Endangered American Wilderness Act and "RARE-II"

Congressional and citizen complaints about the treatment of roadless areas culminated in the successful campaign to enact the Endangered American Wilderness Act of 1978. Launched amid the presidential campaign of 1976 and sponsored by two Democratic presidential hopefuls who were both longtime champions for wilderness, Senator Frank Church of Idaho and Representative Morris Udall of Arizona, this legislation packaged together several dozen long-standing citizen de facto wilderness proposals. Endorsed by candidate Jimmy Carter during the campaign, it became part of his legislative program when he became president.

Hearings on this legislation provided a forum for airing citizen protests about the weaknesses of the RARE process and the flawed Forest Service land-planning efforts to evaluate wilderness. In approving the bill, the House Interior Committee used its committee report to lay out congressional "guidance as to how the Wilderness Act should now be interpreted," directing the Forest Service to correct its stubbornly held purity misinterpretations (direction Congress had initially emphasized in enacting the 1975 Eastern Wilderness Areas Act).

The hearings also led Carter's assistant secretary of agriculture, Rupert Cutler (former assistant to Stewart Brandborg at The Wilderness Society), to announce a new Forest Service effort, which came to be called RARE-II (and thus the earlier process became known as RARE-I). The goal was to better inventory and study national forest de facto wilderness. The RARE-II inventory was a big improvement over RARE-I. This time, with the purity theory officially abandoned, it included roadless areas on national forests in the eastern half of the country. Nearly 3,000 roadless areas were identified and mapped, totaling 62 million acres. However, there were still problems in how the Forest Service went about evaluating the roadless

Two of the greatest congressional wilderness champions,
Senator Frank Church (left) and Representative Morris
K. Udall, watch as President Jimmy Carter signed the
Endangered American Wilderness Act, February 24,
1978 (with this book's author peeking around Udall to
see, too). The law protected seventeen citizen-proposed
national forest wilderness areas totaling 1.3 million
acres. (Courtesy of the White House)

areas. Many activists across the country were dissatisfied with the final 1979
decision, which again sought to sort out some de facto wilderness areas to be made
available for development.

The RARE-II decisions had the effect of impelling congressional action on
millions of acres of de facto wilderness. The focus of debate was not those areas
Assistant Secretary Cutler and the Forest Service recommended for wilderness, but on
the 47 million acres they left in "nonwilderness" and "further planning" categories.
Again, a court decision—this time in a lawsuit filed by the State of California—
settled the matter: the decision (which was upheld on appeal) required that a site-
specific environmental impact statement be completed before each of the inventoried
national forest roadless areas could be developed.

Release Language and the Post–RARE-II Wilderness Laws

The court order in the State of California lawsuit generated intense political
pressures from logging lobbyists and other development interests, who pressed
Congress to find some way to get national forest roadless areas "released" from the
court's requirement for intensive area-by-area review of wilderness potential. While
there was never a similar injunction in any other state and the appeals court ruling
only had force within the states of the Ninth Circuit, the threat was that an identical
lawsuit might be filed and won, affecting other states. Ultimately, the logging
industry and their allies put their political clout behind bills to designate some of

these roadless areas as wilderness, if only to get as much of the other roadless land as they could released and made available for possible roading and timber sales.

Conservationists and their congressional champions—Democratic congressmen Mo Udall, Phillip Burton of California, John F. Seiberling of Ohio, Teno Roncalio of Wyoming, and James Weaver of Oregon—used this industry pressure as leverage. They insisted that any bills that would release some roadless lands from the court's requirement should also designate a substantial portion of the roadless lands as wilderness, a process in which the Sierra Club's chief wilderness lobbyist of that period, Tim Mahoney, was the key strategist for environmental groups.

In my role at the time as conservation director of the Sierra Club, I joined Tim to represent wilderness advocates, at the request of Representatives Burton and Seiberling, in behind-the-scenes negotiations they convened to work out "release language" necessitated by the California lawsuit injunction. Also involved were the chief of the Forest Service and lobbyists for the timber industry coalition. The wording we agreed on had the effect of avoiding California-style injunctions in each state that was subject to a new state-specific wilderness bill. This temporarily sidestepped the site-specific EIS requirement for certain specified portions of the RARE-II roadless lands—as part of a package with new wilderness designations. Because it was so eagerly sought by the timber lobbyists and their allies, this release element had the effect of creating new political momentum behind national forest wilderness bills. The formula allowed wilderness leaders in each state to negotiate which roadless lands would be protected as wilderness and which would be released.

In short order Congress passed national forest wilderness-and-release package bills for many states. In 1984 alone Congress enacted such laws designating more than 8 million acres of new national forest wilderness areas in eighteen states. Most of these laws designated substantially more wilderness than had been recommended in the Forest Service's RARE-II decision. As a result, President Ronald Reagan signed more wilderness laws than any other president—forty-three laws designating 10.6 million acres of wilderness in thirty-one states.

It is important to stress that the release language we worked out only released specified roadless lands for the duration of one cycle of national forest management planning, normally between ten and fifteen years. In fact, we insisted that the standard release language itself expressly require that *all* potential wilderness lands in each state would get a new, fair look by the Forest Service each time the forest plan for each national forest is revised. That plan revision process has been underway for several years now. In the first several dozen national forest plans on which final decisions have been made, the Forest Service has recommended nearly 3 million acres of new wilderness areas.

PART 2: THE BUREAU OF LAND MANAGEMENT JOINS THE WILDERNESS PROGRAM

In a 1926 article Robert Sterling Yard noted, "there are other wildernesses than those in the National Parks and Forests. In the Public Lands, which still have greater area than the National Forests, will be found wilderness regions of charm and beauty."[10] Bob Marshall also advocated wilderness protection on these public domain lands in the 1930s.

From the earliest thinking about the Wilderness Bill, it had been part of Howard Zahniser's ambition to include wilderness-quality lands in the western public domain administered by the Bureau of Land Management. In all, the BLM administers more than 260 million acres of public lands. In his drafts of the Wilderness Bill, Zahniser listed "other public lands" as potentially to be included in the National Wilderness Preservation System. However, the BLM had no established wilderness program and had not administratively designated any areas equivalent to the national forest primitive areas. That made it prohibitively difficult to require a formal wilderness review for BLM-administered lands. There would have been no way to circumscribe the extent of initial wilderness reviews as there was for the 5 million acres of agency-established primitive areas in national forests. Suggesting an inventory of all BLM roadless lands in that era would have been politically unacceptable.

Therefore, as introduced in 1956 the Wilderness Bill simply asserted that the wilderness system could include "such units as Congress may designate by statute," which might include BLM-administered areas at some time in the future. In his 1962 House testimony on the Wilderness Bill, Zahniser said only, "perhaps the Bureau of Land Management may later find that some of the public domain under its jurisdiction is best suited for wilderness preservation."[11]

Sixteen days after he signed the Wilderness Act, President Lyndon Johnson signed a temporary public land–management law governing BLM-administered lands generally. That bill gave the agency explicit authority to administratively set aside lands to protect wilderness values. This only made it more obviously illogical that tens of millions of acres of roadless areas administered by the BLM had been excluded from even study for possible addition to the National Wilderness Preservation System. Wilderness advocates were watching for the right political circumstances to arise in order to correct this glaring gap in the basic architecture of the Wilderness Act.

The Federal Land Policy and Management Act

The opportunity to bring the BLM under the Wilderness Act mandate arose when commercial interests and their friends in Congress sought sweeping "organic" legislation to modernize the legal regime governing these long-neglected public domain lands. This drive had its impetus with Chairman Wayne Aspinall, who was promoting legislation to establish a Public Land Law Review Commission (PLLRC), even as he was blocking the Wilderness Bill in 1962. Aspinall made support for the PLLRC bill part of his price for ultimately agreeing to the Wilderness Bill.

Aspinall himself served as chairman of the PLLRC, which eventually recommended sweeping new legislative authorities for the BLM. Subsequent congressional consideration of this recommendation led to enactment of the Federal Land Policy and Management Act (FLPMA) in 1976. Wilderness advocates, led by Harry Crandell of The Wilderness Society and Sierra Club wilderness lobbyists John McComb and Chuck Clusen, convinced Congress to include a wilderness study requirement. Perhaps because it so obviously corrected a glaring gap in the logic of the Wilderness Act, this wilderness provision was noncontroversial.

FLPMA required the BLM to inventory all roadless areas within its jurisdiction. On the basis of their state by state inventories, which missed perhaps half of the lands conservationists felt should have been included, the secretary of the interior classified some 27 million acres as "wilderness study areas." As specified in FLPMA, these areas received interim protection until Congress could complete action on wilderness designations. Congress has designated some of these areas, but progress has been slow, for much of this BLM roadless land is in states or parts of states that have had congressional representation hostile to statutory wilderness protection, states such as Utah and Wyoming. However, spirited citizen coalitions are at work everywhere, educating and organizing to build grassroots support to encourage Congress to protect more BLM-administered wilderness.

As one of the fruits of this continuing advocacy, in 1994 Congress capped a decades-long citizen campaign by enacting the California Desert Protection Act. This one law established sixty-nine BLM-administered wilderness areas totaling some 3.5 million acres. It also established another 326,000 acres in eight wilderness study areas and expanded Death Valley and Joshua Tree National Monuments, renaming them as national parks and designating large wilderness areas within them.

Nonetheless, the BLM remains the neglected stepchild in the federal wilderness program. While it manages 42 percent of all federal lands, the BLM administers just 6 percent of the wilderness lands protected in the National Wilderness Preservation System—just 6.5 million acres. In what is nothing less than a national disgrace, of the 23 million acres of BLM-administered land in Utah—including some of the most spectacular wild lands anywhere—less than 28,000 acres have been protected as statutory wilderness, not an acre of that at the initiative of Utah's congressional delegation.

At present many of the States vie with one another in boasting of the number of miles of improved roads which they possess. ... It seems worth suggesting that some of the States might well compete to discover which can maintain the finest examples of that glorious frontier which in many ways has been the most distinguishing feature of American history.

Perhaps it may be a better boast for Utah that she still retains the largest roadless area remaining in the United States than that she has more miles of State road than Arizona or Wyoming. Perhaps it may be more honor for Idaho, and incidentally less extravagance, to boast that she has two of the six largest roadless forest areas in the country than to boast of a new highway down the Salmon or the Lochsa River.

It all depends on the viewpoint, but there are multitudes of wilderness lovers and of taxpayers who would delight in being citizens of States which take some pride in preserving the last vestiges of their frontier.

—Robert Marshall,
"A Plea for the Old Wilderness,"
New York Times Magazine, April 25, 1937

Many dedicated officials within the BLM want the agency to become a full partner in the wilderness program, with areas preserved on a scale to match the spectacular wilderness qualities found on the agency's vast land holdings. During the Clinton administration, the agency responded to citizens who pointed out, with detailed field studies, that the BLM had failed to correctly inventory many roadless areas, notably 2.6 million acres of roadless lands in Utah. Those areas became officially recognized for further review. And, in a way paralleling the recent practice of the Forest Service, BLM field offices used their planning procedures as an opportunity to both correct inventory errors of that sort and to select additional wilderness study areas for recommendation to Congress.

Unfortunately, these positive steps were halted in 2003 by Secretary of the Interior Gale Norton by means of her secret settlement of a lawsuit brought by the State of Utah. Her decision is being contested in federal court, but as I wrote this in the summer of 2004 the BLM was at least temporarily precluded from recommending *any* additional wilderness preservation to Congress beyond the 15 million acres of wilderness study areas identified in the 1980s pursuant to the initial inventory required by FLMPA but not yet resolved by Congress. Clearly, there are more chapters yet to unfold in the come-from-behind wilderness program of the Bureau of Land Management.

The Loss of Big Wilderness

In the last decades of the twentieth century, real progress was made in gaining statutory wilderness protection for more federal lands administered by the U.S.

Forest Service and Bureau of Land Management. That progress came at a pace few might have predicted with any confidence in 1964.

It must be noted, however, that all through the history recounted and celebrated here, there were major losses of de facto wilderness lands as well. The sad reality is that the relentless pressure of a seemingly insatiable tide of road building and development accelerated in these decades. Despite every effort, no comprehensive way had been found to wholly stop the steady loss of national forest roadless lands to logging. The same was true for BLM-administered lands under pressures for oil and gas leasing, mining, and other development.

The greatest loss was in the special quality inherent in large wilderness expanses, for in too many cases roadless areas have been dismembered into smaller and smaller fragments by new roads. Deep-forested valleys were roaded to their very upper reaches, in many cases just to preclude later wilderness designation of these valleys. The loss of big wilderness is masked in acreage statistics, for larger roadless areas were fragmented while still leaving much of the land roadless, but in many smaller units. Aldo Leopold called it "the process of splitting."[12]

The regrettable historical reality is that the movement to protect wilderness and to obtain official wilderness protection policies had to be built up through most of the twentieth century. Progress was often painfully slow, even as roadless lands were falling to chainsaws and bulldozers. As Aldo Leopold warned in 1925:

> An incredible number of complications and obstacles ... arise from the fact that the wilderness idea was born after, rather than before, the normal course of commercial development had begun. The existence of these complications is nobody's fault. But it will be everybody's fault if they do not serve as a warning against delaying immediate inauguration of a comprehensive system of wilderness areas.[13]

For my own part, I strongly believe future generations will judge that most of these losses of big wilderness were wrongheaded, fueled by short-term thinking and a parochial rush to sustain unsustainable levels of logging and other development at the price of a long-term need for more wilderness.

Despite disappointments and temporary setbacks, important progress is being made to protect a great deal of wilderness on these most vulnerable categories of federal land. The fact that the Forest Service undertook the RARE-I and RARE-II inventories in the 1970s led agency officials to systematically identify roadless lands for the first time. These agency inventories, flawed though they were, have given both local wilderness-minded agency personnel and activist groups new tools, notably helping citizen groups focus on potential wilderness areas they had not previously studied. Agency and citizen inventories of these roadless areas established them as a political fact of life, even as their legal status often remained subject to contentious debate.

The National Forest Roadless Area Conservation Rule

During the 1990s President Bill Clinton ordered development of a formal government rule to protect all national forest roadless lands. Formal adoption of the rule included in-depth public comment opportunities addressing the areas the agency had inventoried in the RARE-II process. On the basis of 600 public hearings nationwide and the largest outpouring of public comment in the history of all federal rule-making, the secretary of agriculture adopted the Roadless Area Conservation Rule just before the Clinton administration left office in 2001. The rule gave a new form of high-level protection to the 58 million acres of national forest roadless areas identified in the RARE-II process.[14]

When the George W. Bush administration took office, it sought ways to reverse the roadless rule. The rule was also challenged in numerous court cases, and for the most part the federal government sat on its hands, not even appearing in court to defend the roadless area protection. At the same time, the Bush administration worked to lift or weaken the new protections. Over intense public protest, it exempted the heavily threatened roadless areas of the Tongass National Forest from protection under the rule.

In June 2004 the Bush administration political appointees overseeing the Forest Service announced a proposed replacement for the roadless rule that would give the governor of each state unprecedented authority to intrude in the protective provisions for national forest roadless areas. The administration explained its proposal:

> Due to the continued legal uncertainty of providing protection for roadless areas through the application of the roadless rule, the agency is proposing to amend the roadless rule by replacing the prohibitions of the January 2001 rule with a procedural rule that would set out an administrative process for State Governors to petition the Secretary of Agriculture to establish or adjust management direction for roadless areas within their State.[15]

This proposal would eliminate protections afforded to these last wild places by the roadless rule, leaving the crucial land-use decisions for these *federal* lands in the hands of select *state* politicians.

Whatever unfolds next, at the very least roadless areas have achieved a well-recognized status in the real world, valued as important natural resources in their own right, with thousands of citizens rallying to protect them. With the power of the National Environmental Policy Act, the sometimes secretive planning for federal lands has been opened up. As never before, plans to develop roadless areas for development are something local people know about and can more effectively challenge in the courts. They can also appeal directly to Congress to advocate wilderness protection if they choose.

PART 3: WILDERNESS ON A SCALE TO MATCH ALASKA

The greatest expansion of the wilderness system came in 1980 with enactment of the single largest land conservation measure in the history of our or any
other nation.

Alaskan wilderness had fired the imaginations of Americans from before the
time of John Muir. For decades pioneer conservationists in Alaska sought stronger
protection for the unsurpassed wilderness of "The Great Land." However, prior to the
Wilderness Act, the Forest Service had designated no wilderness or primitive areas
administratively in the two huge national forests in southeast Alaska. Therefore, no
areas in Alaska came into the National Wilderness Preservation System or were
required to be studied in 1964 act. However, the new law did require wilderness
reviews for the then-existing Alaskan national parks and national wildlife refuges.

Not daunted by the lack of foresight by Forest Service officials, conservationists
in southeast Alaska prepared their own citizen wilderness proposals for large,
spectacular forest roadless lands, but a generally hostile congressional delegation made
approval seem a distant dream. Most attention had to be given to the constant fight
to turn back massive, heavily subsidized timber sales on the Tongass National Forest.

During the 1960s and 1970s, land-use decisions across the vast expanse of
federal lands in Alaska were bound up in larger historic forces, notably the pressures
to make federal lands available for wide-open development of all kinds, unresolved
aboriginal land claims, and the 1968 discovery of oil at Prudhoe Bay. When Alaska
Native organizations asserted their land claims, Interior Secretary Stewart Udall
imposed a land freeze in 1966, stopping land transfers including those to the State of

---

There are glaciers, mountains, fjords elsewhere, but nowhere else on earth is there such
abundance and magnificence of mountain, fjord and glacier scenery. For thousands of
miles the coast is a continuous panorama. For one Yosemite of California, Alaska has
hundreds. The mountains and glaciers of the Cascade Range are duplicated and a
thousandfold exceeded in Alaska. ... Its grandeur is more valuable than the gold or the
fish, or the timber, for it will never been exhausted. This value, measured by direct
returns in money received from tourists, will be enormous; measured by health and
pleasure, it will be incalculable.

—Henry Gannett, in *Harriman Alaska Expedition*,
quoted in Natural Resources Defense Council booklet
"Celebrating Wild Alaska: Twenty Years of the Alaska Lands Act"

Alaska. This conflicted with the rush to develop oil fields, construct the Trans-Alaska Pipeline, and other development ambitions. The Udall land freeze had to be lifted in order to get the oil flowing and additional federal land transferred to the state. The pent-up demands for federal land transfers and development could only occur when Congress resolved the Native land claims issue.

The upshot was that in the late 1960s great pressure existed from the state, the oil industry, the Nixon administration, and developers of all sorts to pass the proposed Alaska Native Claims Settlement Act, a bill to implement a settlement that had been negotiated with Native organizations.

A few conservationists in Alaska and lower forty-eight leaders, notably Harry Crandell of The Wilderness Society and Dr. Edgar Wayburn of the Sierra Club, realized that when the land freeze was removed, the slicing up of Alaska lands—375 million acres of Alaska almost entirely then owned by the American people—would surpass the frenzy of the nineteenth-century Oklahoma land rush. In the rush to transfer millions of acres of lands out of the federal estate, would anyone speak up for the national interest in reserving some great expanses in Alaska as national parks, national wildlife refuges, wild and scenic rivers, and wilderness?

The Alaska Coalition and "D-2"

Alaska conservation groups and The Wilderness Society and Sierra Club formed the core of the Alaska Coalition, which they established in 1970. In a come-from-behind effort in which I was a leader, the coalition and other groups urged Congress to include in the proposed Alaska Native Claims Settlement Act a provision to require that some large expanses of federal land be selected and held in federal ownership, protected from development or transfer to the state pending more detailed study of their potential as new national parks, wildlife refuges, and wilderness areas.

The Alaska Coalition pushed this amendment to a vote on the House floor in 1970. The Udall-Saylor National Interest Lands Amendment was defeated, but the impressive size of the favorable vote, against formidable forces lobbying against us, helped create political momentum for the "national interest lands" idea. As a result, building on a similar idea in the Senate bill, a national interest lands provision was ultimately incorporated in the Alaska Native Claims Settlement Act signed by President Richard Nixon in 1971.

This "D-2" provision—so-called because it was section 17(d)(2) of the law—gave interim

> Alaska is the only place left in the United States where it is still possible to experience wilderness on a scale comparable in its vastness, splendor and pristine purity to that which astonished and dazzled the early explorers of America.
>
> —John F. Seiberling,
> in the introduction to *Alaska, Wilderness Frontier* by Boyd Norton

legal protection to certain federal lands in Alaska selected by the secretary of the interior as having highest values for park, refuge, and wilderness purposes. This allowed Congress time to act on the results of more detailed studies. The interim protection would last until December 1978.

During the mid-1970s, the federal agencies undertook extensive land-use and wildlife studies across Alaska to fulfill the D-2 mandate. These studies led to a decision by the administration of President Gerald Ford to give interim protection to some 82 million acres of potential "national interest lands." In parallel with those efforts, Alaskan and national conservation groups developed their own comprehensive citizen proposals for new and expanded national parks, wildlife refuges, wild rivers, and wilderness areas designed on a scale to match Alaska's incomparable wilderness and wildlife resources. The question became how to get this unprecedented set of proposals enacted into law before the end of 1978.

Conservation leaders realized that such a visionary program would draw powerful and united opposition from Alaska boosters, the Alaska congressional delegation, and every conceivable development interest from oil to mining. To get their package through Congress would require a far larger, more cohesive effort than anything America's conservation movement had ever mounted. With the gifted leadership of its chairman, Chuck Clusen of the Sierra Club, the Alaska Coalition united Alaskan and national conservation groups and amassed unprecedented resources for this campaign. It was my good fortune that Chuck—my graduate school roommate and later my colleague on the staff of the Sierra Club—asked me to design and coordinate the lobbying element of the Alaska Coalition's unprecedented grassroots, media, and lobbying campaign.

The Alaska National Interest Lands Conservation Act

In 1977, with our citizen-developed Alaska National Interest Lands Conservation Act (ANILCA) introduced in the Congress, an extraordinary constellation of leaders came together behind it:

- President Jimmy Carter came into office, with Cecil Andrus as his secretary of the interior. Both were fired by the vision of a truly historic Alaska Lands law.
- Representative Mo Udall, sponsor of the original "national interest lands" amendment back in 1970, became chairman of the House Committee on Interior and Insular Affairs and the lead sponsor of the legislation. He gave top priority to the Alaska lands issue.
- Representative John Seiberling became chairman of Udall's special Alaska lands subcommittee. Seiberling was an essential leader, teaming his extraordinary command of the complexities of the issues and the Alaskan landscape with Udall's unmatched legislative skills. Both were backed by an expert staff, headed by Harry Crandell.

The campaign to pass ANILCA—the most extensive and cohesive campaign in the history of the environmental movement—also brought out a formidable host of opponents. The Alaska lands issue raged through the late 1970s. We triumphed in passing a very strong bill in the House in 1978 by an overwhelming nine-to-one vote margin. When a Senate filibuster killed the legislation, and with the interim protection of the D-2 lands set to expire that December, President Carter and Secretary Andrus stepped in to protect the land by administrative order. This spurred even the bill's staunchest opponents to want the next Congress to act, if only to be able to fine-tune the protections and boundaries set in the presidential and secretarial orders.

In 1979 the House again passed a very strong bill, but the Senate committee approved a much-weakened and less expansive version. When President Carter and our coalition won the first of a projected series of Senate floor votes on amendments to expand this committee bill, Senate leaders pulled it from further floor action. Instead, they convened behind-closed-door negotiations dominated by those opposing the House-passed Udall-Seiberling bill.

The Senate bill that resulted from these one-sided negotiations was still unsatisfactory to conservationists. Congressmen Udall, Seiberling, and Republican Tom Evans of Delaware prepared a much stronger substitute version they hoped to have the House adopt and return to the Senate in order to up the ante in final negotiations. However, the 1980 election intervened and Ronald Reagan defeated President Carter. It was obvious that as president, Reagan would side with opponents of the Alaska lands bill. So, during the post-election lame-duck session, House leaders and conservationists had no real choice but to accept the less expansive but still historic Senate version.

President Jimmy Carter signed the Alaska National Interest Lands Conservation Act on December 2, 1980. In addition to establishing vast new national parks and wildlife refuges, the act designated some 55 million acres of wilderness, more than doubling the size of the National Wilderness Preservation System the moment the president lifted his pen from the new law. New wilderness areas were designated in existing and new national parks (32 million acres in eight park units) and existing and new national wildlife refuges (18 million acres in thirteen refuges). In a particularly hard-fought victory, long-stalled wilderness protection was provided at last for citizen-proposed national forest wilderness areas in southeast Alaska (5 million acres in fourteen wilderness areas).[16]

The 1980 Alaska lands law set the stage for the

As long as any of us serve in this House, we will cast no more important conservation votes than the votes on this Alaska lands bill. The simple fact is that Alaska ... is our last chance to go about the job of conservation ... with forethought, before a pattern of development has been fixed across the landscape piecemeal.

—Representative Morris K. Udall,
Congressional Record,
April 23, 1979

next chapters in the saga of preserving Alaska Wilderness:

- With tens of millions of acres of wild lands under its care in Alaska, the Bureau of Land Management has been prohibited from conducting any wilderness studies or considering recommendation of statutory wilderness during its land-use planning. Similarly, the National Park Service and the Fish and Wildlife Service have been discouraged—or outright forbidden—from completing wilderness studies and making recommendations in the portions of their Alaskan areas not yet protected as wilderness.

- While important wilderness areas were designated in the Tongass National Forest, many other world-class wild places there remain threatened by the grotesquely subsidized logging industry. In 1990 Congress passed and President George H. W. Bush signed the Tongass Timber Reform Act, an important partial reform of the logging subsidies and unsustainably high timber-cutting targets. This law also designated an additional 296,000 acres of wilderness and placed mandatory protections over another 730,000 acres of roadless areas to be retained with their wild character.[17] The struggle over logging, road building, and the fate of national forest roadless areas in Alaska continues to this day.

- Though the House-passed Alaska lands bill had designated the entire coastal plain of the Arctic National Wildlife Refuge as wilderness, the final law excluded 1.5 million acres from this protection. However, it did specify that commercial oil exploration and development would require approval in a subsequent act of Congress, shifting the burden to the oil lobbyists. Despite intensive lobbying by the Bush-Cheney administration and its powerful allies, conservationists have been able to rally public support to block all efforts to pass such an oil-drilling bill.

Nonetheless, the present political impasse will eventual crumble, and I am utterly confident in predicting that much more wilderness will receive Wilderness Act protections across the wild wonders of Alaska. On its tenth anniversary, the late T. H. Watkins wrote that the 1980 Alaska Lands Act

> was at once one of the noblest and most comprehensive legislative acts in American history, because, with the scratch of the presidential pen that signed it, the act set aside more wild country than had been preserved anywhere in the world up to that time—104.3 million acres. By itself, the Alaska Lands Act stood as a ringing validation of the best of what the conservation movement had stood for in the century since Henry David Thoreau had walked so thoughtfully in the woods of Walden Pond.[18]

Watkins captures the scale and continuing opportunity that so strongly motivates wilderness advocates everywhere to persist in their efforts to protect more of the world-class wilderness heritage of "The Great Land."

President Jimmy Carter and his wife, Rosalynn, relax on the coastal plain of the Arctic National Wildlife Refuge with Alaska wilderness leader Debbie Miller and her daughter Casey. On this 1990 trip, the Carters closely observed musk oxen and, at one point, were surrounded by 80,000 caribou. It was, Carter recalled, "a profoundly humbling experience. We were reminded of our human dependence of the natural world." (Courtesy of Debbie Miller; Dennis Miller, photographer)

Chapter 6

Wilderness Yet to Save

Through the history I have traced, many area of America's wilderness heritage are now firmly protected by law in the National Wilderness Preservation System. But there is more—much more—wild federal land that remains unprotected and inadequately studied. Much of this is land neither the agencies nor Congress has yet considered in a sufficiently detailed, case-by-case review.

Our Unprotected Wilderness

Many millions of acres of public land are still largely unroaded but have no statutory protection to ensure they will remain part of our wild bequest to future generations. Millions of Americans use these unprotected areas now *as wilderness*, not differentiating them from those "capital-w" wilderness areas already well-protected thanks to the Wilderness Act. The wilderness qualities of these lands remain at serious risk, threatened by the same old nibbling away by piecemeal development decisions.

As David Brower told a 1962 Seattle wilderness conference, these unprotected, still-wild federal lands "aren't legally wilderness areas at all. They're just plain wild. They are de facto wilderness. In my favorite definition, they are simply 'wilderness areas which have been set aside by God but which have not yet been created by the Forest Service.'" "De facto wilderness," he added, is "the wilderness that sits in death row ... and there has been nothing ... like ... a fair trial."[1]

What happens as more and more of our remaining roadless lands are devoured? Obviously, the natural values of the lands are compromised and their solitude defiled. But there is a subtler impact as well, for those who have loved and used these places as lowercase-w wilderness are then displaced, and they must then concentrate their use on an ever smaller fraction of the still-wild public lands that are left.

No one I know asserts that every acre of this roadless land will be or should be designated as wilderness. But surely we should take care to think seriously about each area before we subject it to our vast powers to manipulate, deform, and shape the land to our own will. We should demand no less from our federal agencies, for we should all think—and think long and hard—in our role as agents for those who will live with our choices in the twenty-third century and beyond.

The threats to these lands are all the usual ones—oil and gas drilling driven by our grotesquely unsustainable energy policy of using up fossil resources as rapidly as we possibly can; chopping down our old-growth forests and wild second-growth forests, too; invading everywhere with motors and machinery. All of these threats invade de facto wilderness in the same way—with roads. Roads kill wilderness, and not only by introducing our destructive mechanical accomplices. Roads also kill wilderness by a death of a thousand cuts. A large wilderness is sliced in two by a road. In the statistics, the roadless acres seem little changed, for the roadbed itself takes up a very small portion. But in reality we have lost the expansive scale of another wild place as we slice and dice large areas into ever smaller fragments. Aldo Leopold wrote to Bob Marshall in 1937, concerned about projected roads in the largest remaining national forest roadless area in Missouri, "it looks like another case of chopping up a wild area and then labeling one of the chips a wilderness."[2]

That same year, Marshall wrote in the *New York Times Magazine* that "there are nearly 3,500,000 miles of roads" in the United States, and that "road systems have been expanding so rapidly that few people have had time to stop and think."[3] Imagine his reaction were he here to see that America now has doubled its road mileage to 7 million miles.

> Get into your car and start driving. Set the cruise control at 50 miles per hour. Do not stop for food or bathroom breaks, do not stop for gas, do not stop for any reason. Drive 24 hours per day without rest until you have covered every official mile of road in the United States. You'll be done in 16 years with your odometer reading roughly 7,000,000.
>
> —Joshua Brown,
> *Wild Earth*, Summer 2002

How Much Unprotected Wilderness?

An analysis for the Campaign for America's Wilderness in 2001 found some 319 million acres of the federal lands administered by the Forest Service and Bureau of Land Management still unroaded.[4] These wild lands outside presently designated wilderness represent just 14 percent of our nation's landmass—and none of them have statutory protection to assure they will remain a wild legacy for future generations. Consider these statistics:

- Together, the Forest Service and BLM administer 456 million acres of our federal lands. Of this, some 360 million acres (including designated wilderness) were unroaded at the time of our 2001 survey.
- Of these unroaded federal lands, only about 41 million acres are now designated as capital-w wilderness. That leaves 316 million acres we have called *unprotected* wilderness.
- Because so much of our now-designated wilderness is in Alaska, it is important to examine the situation of the unprotected wilderness in the rest of the country:
  - Outside of Alaska, some 47 million acres of federal land (in all agency jurisdictions, not just Forest Service and Bureau of Land Management) are preserved as wilderness. This amounts to just 2.5 percent of the landmass outside of Alaska, in all ownerships.

- Outside of Alaska, the Forest Service and BLM administer some 348 million acres. Of this, some 257 million acres remain unroaded.
- Of these unroaded federal lands outside Alaska, some 35 million acres are now designated as capital-w wilderness. That leaves 222 million acres of unprotected wilderness.

The upshot: seven out of every eight acres of these wild federal lands that Americans now use and cherish as wilderness are not protected by law.

These are big numbers. Bear in mind that our inventory identified unroaded lands down to a minimum unit of 1,000 acres. Many of these smaller units will not be protected as wilderness. Nor are all of the larger unprotected wilderness areas likely to be designated by Congress for permanent preservation. But these last still-roadless public lands should not be heedlessly sacrificed to development. Their wild values merit thorough on-the-ground, area-specific study and fair consideration. *Yet despite overwhelming public interest, the vast majority of these lands have never been given thorough and fair consideration for permanent protection as wilderness.*

De facto wilderness is wilderness in fact—a big, wild, unmechanized, unroaded natural place with no formal border around it. It's just there, wild and beautiful, and important or unimportant as wilderness to the Nation's future. We don't know. We just own it.

—David Brower,
"De Facto Wilderness: What is Its Place?"
Wildlands in Our Civilization

Giving Roadless Areas a Fair Shake

For more than thirty years, the question of how to decide the future of the still-roadless lands administered by the U.S. Forest Service and the Bureau of Land Management has dominated wilderness politics. Notwithstanding this long history of controversy, today the issue remains as sharp and hotly contested as ever in many states. For wherever there is a prized roadless expanse of slickrock canyons, old-growth forest, or Alaskan vastness, citizens who know these places are working to defend them from ill-considered development and the invasion of off-road vehicles.

In an ideal world, these decisions would be made calmly and with broad community consensus. But the stakes are such, it seems, that controversy is unavoidable. In a 1962 talk, Dave Brower assessed these as

decisions that do not have to be made in a hurry and they should not be. They are hard decisions in proportion to the effort to hurry them. They are decisions that no man again in America will ever have the chance to unmake—they are as irrevocable as divine law; but I fear that they are being made with something less than divine judgment.[5]

The way this *ought* to work, it seems to me, is that the federal agencies should conduct scrupulously fair and open planning procedures in which the future of each roadless area is thoroughly reviewed. Citizens both locally and nationally should have equal opportunity to be heard—and listened to. And the agencies should be able to make their wilderness recommendations (or their recommendation not to designate) without undue political interference from whichever party may hold the White House and cabinet offices. The cabinet secretary involved, or the president, should make their recommendations. And then all should have an opportunity to present their views in the one place where the fate of wilderness is supposed to be decided—in Congress. Congress should decide case by case.

In 2004 the evidence shows that the Forest Service is doing the best job among agencies. Congress has required that each time a national forest management plan goes through its required periodic revision, at least every fifteen years or so, the agency "shall review the wilderness option," giving any still-roadless lands a new look, and shall specifically evaluate, with opportunities for public comment, the possibility of recommending some or all to Congress for designation as wilderness.[6] This puts the question of new wilderness designation on the table for open public discussion.

The 2003 revision of the plan for the Wasatch-Cache National Forest, in the mountains rising literally on the eastern edge of the Salt Lake City urban area, offers an example. As the regional forester's decision (which was reviewed in advance by Bush administration political appointees) explains

"the previous Forest Plan ... reflected the desires that the public had nearly 18 years ago when the primary focus was on what the land could produce. These desires have changed, and they will continue to change. Today's focus is centered more on the condition of the land as a basis for providing multiple goods and services. ... Rather than allow development in most roadless areas as the 1985 Plan did, the Revised Plan maintains roadless area values in 75 percent of Inventoried Roadless Areas (IRAs) and recommends about 73,500 acres for wilderness designation."[7]

The regional forester went on to explain

Public opinions on the issue of Wilderness varied widely during the planning process. There was much support for recommendations of large amounts of new Wilderness across broad areas of the Forest on the Wasatch Front and in the Uinta Mountains. Wilderness recommendations were considered as important to maintaining ecosystems, and placing limits on development and motorized access. Many others wanted no new Wilderness recommendations anywhere on the Forest and were usually concerned about the areas where they had existing permitted uses or motorized access, or simply enjoyed the existing management settings and wanted no changes.[8]

On the basis of this new look and its consideration of public comment, the Forest Service recommended several new wilderness areas and important small additions to others—on a national forest where 25 percent of the land is already designated wilderness. Of course, here and elsewhere, local citizen wilderness groups may wish to see more land designated as wilderness. If so, the Wilderness Act gives them the means and the avenue. They, like the agency, may take their own proposals directly to Congress. Those who oppose wilderness designation have exactly the same opportunity.

Giving Roadless Areas the Brush-Off

The Forest Service approach, as directed by Congress, allows for an open review of the future of roadless lands in the context of agency land-use planning. This may seem so obviously sensible as to hardly merit praise. But I praise it because it contrasts so dramatically with the hostility that currently exists from Bush-Cheney appointees in the Department of the Interior. As a result of this hostility, the administration has single-handedly created an outlook of deepening public controversy for decisions about roadless lands administered by the Bureau of Land Management. Within days of the Wasatch-Cache forest plan decision recited here, and in the same state, Secretary of the Interior Gale Norton chose to conclude a secret and swift negotiation with the governor in which she officially disclaimed any authority by the Bureau of Land Management *to even consider* the option of recommending new wilderness areas to Congress during the agency planning process. Her action had immediate application to all BLM-administered roadless lands nationwide.

The wilderness issue has been unusually polarized in Utah, where the majority of the congressional delegation has long blocked new wilderness designation for BLM-administered lands. The BLM has less than 28,000 acres of statutory wilderness

Defenders of the short-sighted men who in their greed and selfishness will, if permitted, rob our country of half its charm by their reckless extermination of all useful and beautiful wild things sometimes seek to champion them by saying "the game belongs to the people." So it does; and not merely to the people now alive, but to the unborn people. The "greatest good for the greatest number" applies to the number within the womb of time, compared to which those now alive form but an insignificant fraction. Our duty to the whole, including the unborn generations, bids us restrain an unprincipled present-day minority from wasting the heritage of these unborn generations.

—Theodore Roosevelt,
A Book-Lover's Holidays in the Open

in Utah, out of 23 million acres they manage. By contrast, a citizen bill developed to defend against incessant efforts by Utah politicians to legally write-off potential wilderness areas proposes to protect more than 9 million acres of the stunningly wild, scenic redrock wilderness across southern Utah.

The political maneuvering by Secretary Norton and former Utah governor Mike Leavitt (later appointed to head the Environmental Protection Agency by President Bush) sets a new standard for cynicism. Leavitt filed a new version of a dormant lawsuit, which the state had lost on appeal. It took Secretary Norton just two weeks to closet herself with Leavitt and sign a closed-door settlement, capitulating to the state in every detail without even a thin pretense of defending the federal interest in court.

Secretary Norton ordered the Bureau of Land Management to abandon its orderly process in which agency recommendations for wilderness protection are developed on the factual basis of on-the-ground planning with full opportunity for public participation. And her order applied nationwide. Thus, while the Forest Service is able to take into account new information and changing public desires as it revises its management plans, in just one particular the BLM is forbidden to do so—and that is if the question involves wilderness.

Anything but Statutory Wilderness

What Secretary Norton and the forces behind her hate—and *hate* is not too strong a word—is the *statutory-ness* of wilderness designation, the presumption of perpetuity, which is, of course, the very point of the Wilderness Act. In this they have been even more blatant than other wilderness opponents. Knowing that the American people like wilderness protection and overwhelmingly support protecting more (as demonstrated in poll after poll), Secretary Norton attempted to mask the real effect of her decision. In a public-relations effort to disguise what was really being done, her order promises that the BLM may still inventory and consider "characteristics that are associated with the concept of wilderness."[9] What the agency cannot do is recommend any such lands to Congress for statutory protection under the Wilderness Act.

Why get so doctrinaire about the fact that wilderness designation is statutory? By cutting off the route by which the agency could even suggest new wilderness areas to Congress, the goal is to leave any protection of "wilderness characteristics" to purely administrative decisions. These are decisions the political appointees can control and change—the old problem of wild lands protected by the stroke of a bureaucrat's pen and as easily opened again for development. Appointees cannot so easily change boundaries or protections for statutory wilderness areas once they are enacted by Congress. So, this Bush policy constitutes a giant step backward to the pre–Wilderness Act era.

This idea pleases the development and "access" constituents for whom these BLM-administered lands are prime targets, including proponents of accelerating oil and gas leasing and development. A January 2002 Bureau of Land Management memo could not be clearer:

> Wilderness [review] procedures appear to hinder, rather than promote access to oil and gas resources on public lands. This is inconsistent with the administration's energy policy.[10]

Must it always be statutory wilderness or nothing? Do wilderness advocates simply draw a line against any statutory recognition of other kinds of recreation incompatible with wilderness designation? Not at all; the conservation movement has been working for new national parks and recreation areas since the time of John Muir. Mountain cyclists have a fair claim to their preferred use on public lands—and millions upon millions of acres are open to them. The same is true for off-roaders, most of whom are not scofflaws out to trash every last acre of wild land. And the wilderness movement has a record of constructive involvement with the devotees of these forms of recreation.

Take mountain bikes. In numerous cases, local bicycling clubs and local wilderness groups have been able to work out reasonable accommodations so that long-favored bicycle routes are excluded from boundaries for a proposed wilderness area. In some cases, unique protective provisions have been included in a separate element as part of legislation that also designates the wilderness area. Such provisions assure that nonwilderness areas can be better protected against adverse development while allowing bike use. In individual cases such areas adjacent to designated wilderness have been named "national recreation area," "national conservation area," "protection area," or "scenic area."

One example is the proposed Virginia Ridge and Valley Wilderness and National Scenic Areas Act, which was introduced in 2004 with bipartisan support within the Virginia congressional delegation.[11] Local biking groups and the Virginia Wilderness Committee worked out boundary compromises to accommodate everyone's uses while protecting wilderness areas for the far broader, nonrecreational purposes of the Wilderness Act. In a press release, International Mountain Bicycling Association spokesman Gary Sprung emphasized that Virginia wilderness advocates

> worked hard to understand and accommodate mountain bicycling. They made many boundary adjustments and supported National Scenic Areas for the two areas with significant bicycling trails. ... We did give up some trails that were important to bicycling in the proposed Raccoon Branch Wilderness. But we consulted with local cyclists and most agreed that the agreement made sense.[12]

The BlueRibbon Coalition, an off-road vehicle lobbying group, and similar interests are eager to convince us that the Wilderness Act is a straightjacket, and that nonstatutory "protection" through management plans is more flexible and therefore wiser. "We all want those treasures protected," says an advertisement distributed in 2003 and 2004 by these groups, "but wilderness is not the best way." The identical ad has shown up in Idaho (sponsored, seemingly independently, by "Responsible Idahoans") and in Washington state (sponsored by the Washington State Snowmobile Association). The ads seek to block wilderness designation proposed for portions of Idaho's Boulder–White Cloud Mountains, the Owyhee Desert, and the proposed Wild Skykomish wilderness, respectively. "Management under B.L.M. and Forest Service Management Plans have served these lands well and will continue to do so," the two ads assure us:

> Carefully crafted plans that we help develop assure these lands and the ecosystems they support will be sustained for present and future generations. Management emphasis can vary from leaving the land undeveloped to building roads and harvesting trees. Managers have access to a full toolbox of management options that wilderness denies them. They can enhance endangered species habitat and actively protect it. They can prescribe and ignite fires needed for fuel management. They can thin dense stands of trees, salvage dead trees and remove barriers to endangered salmon. They can manage for a variety of recreational experiences.[13]

In other words, the plan is ever flexible ... enough so you can drive a truck through it. Which is their point.

If those who wish to block most or all new wilderness designations cannot head them off completely, they have an alternative which would result in the same thing. The BlueRibbon Coalition and others have been promoting an alternative "backcountry" designation—a sort of "wilderness lite" featuring machines and motors. The idea is to compete with wilderness designation on the national level, but with a new generic legislative designation that would "allow traditional multiple-use activities with the emphasis on promoting and protecting recreation, not systematically eliminating it as in Wilderness."[14] This proposal is all about that catchword *access*—access for off-road vehicles, access for machinery (including mountain bicycles), access for what one proponent calls "'light touch' or compatible management activities," and access for "management activities to protect the forest" and to "improve wildlife habitat."[15]

Defined this way, an alternative "backcountry" status, even if statutory, is not wilderness and not even vaguely like it. Advocates of the backcountry alternative are eager to set up a wilderness-versus-backcountry debate, as if the issue is solely a question of one kind of recreation versus another. But allowing motors and machines, let alone "light touch" management of any sort not permitted by the Wilderness Act, strikes at the absolute heart of the wilderness concept as defined by Aldo Leopold and Bob Marshall. As Dr. James Gillian posed the question in his 1954 scholarly paper on wilderness and democracy, "the democracy of providing access to every parcel of our public lands may triumph, but will future generations appreciate this particular brand of wisdom?"[16]

Bill Dart, the executive director of the BlueRibbon Coalition, announced in early 2004 that they have hired a Washington lobbying firm to work to refine the backcountry idea:

> Some folks in the natural resource industry groups have legitimate concerns about any new federal land designation that might make further restrictions on their industries. It has never been our intention to restrict multiple use land. Our focus has been on lands that have already been determined suitable for Wilderness designation, to provide the back country concept as an alternative to Wilderness, rather than back country being an additive designation for other lands currently used for timber production, mining, grazing, or other multiple uses.[17]

Dart's predecessor, Clark Collins, had earlier introduced the backcountry concept by stressing that the goal is "to protect and if possible enhance the back country recreation opportunities on these lands while still allowing responsible utilization of these areas by the natural resource industries."[18] So, they propose to allow the extractive industries to invade areas already determined to be suitable for wilderness. In another place Collins spelled it out:

> By permitting some low impact resource extraction (i.e., selective timber cuts, helicopter removal, allowing only a few, temporary non-permanent roads), industry would be able to have some commercial access to an area that is becoming politically difficult and would be legally impossible to access if designated "Wilderness." There are many questions to be answered. Some that come to mind are

---

Wilderness holds within itself all the mystery of the universe, the story of evolution, of growth and change and beauty from the beginnings of time. Wilderness is more than lakes, rivers, and timber along the shores, more than fishing or just camping. It is the sense of the primeval, of space, solitude, silence, and the eternal mystery. It is a fragile quality and is destroyed by man and his machines.

—Sigurd Olson,
testimony before the Selke Commission
on the fate of the Boundary Waters Canoe Area,
May 23, 1964

- Can this designation be made acceptable to multiple use recreationists?
- Can it be made acceptable to forest commodity users?
- More importantly, can it be made acceptable to Congress and the general public?

Our goal is to develop a BRA [Backcounty Recreation Area] land classification proposal to the point of obtaining a consensus support from both multiple-use recreation and the resource industries.[19]

The Majority of Americans Are "Not at the Table" as the Fate of Wilderness Is Decided by Agencies

Allocating lands among competing and often-incompatible uses is the heart of land-management planning for federal lands. Federal law mandates the meaningful involvement of the public in these decisions. For example, the required public involvement in planning for the national forests is intended to "ensure that the Forest Service understands the needs, concerns, and values of the public."[20]

Agency planning procedures do commonly include public hearings and solicit written public comments as a means of helping decision-makers understand the concerns and values of the public at large. However, these hearings are almost always held in local communities not convenient to more-distant Americans who are equal owners of these public lands and whose tax dollars often heavily subsidize their development. Certainly, the more broadly such hearings are held and the more widely written input is solicited, the more complete a picture of public "needs, concerns, and values" officials will obtain.

In current agency practice, commercial interests such as oil and gas drilling, logging, and mining are well represented at local decision-making tables. But it has been rare indeed for the federal agencies to use scientific public opinion polling to scientifically establish and assess broader public views as they weigh fundamental decisions for the development or preservation of public lands. As the Forest Service's National Survey on Recreation and the Environment (NSRE) has shown, Americans increasingly give highest importance to the "nonuse" values served by protecting wilderness—benefits such as protecting clean water and wildlife habitat and leaving a legacy of wildness for the future. But as the director of the NSRE, Dr. H. Ken Cordell, and his research colleagues observe, "Usually, non-use interests—that is, the interests of the majority of Americans—are not at the table."[21] As the natural resource planner for the southern region of the Forest Service put it, those attending local public meetings "typically represent only a portion of the public's interests and seldom represent the so-called 'silent majority' who do not or cannot attend these meetings."[22]

As part of its planning process for a number of southern national forests, the agency did take the unusual step of commissioning special polling conducted by the NSRE survey scientists. The 2001–2002 polling surveyed adults in the "primary market area" within seventy-five miles from the boundaries of these forests. Regarding their own participation in various recreational uses within the national forests in the past twelve months, the public responses challenge stereotypes. Of twenty recreational activities, the fourth highest was "visit a wilderness or undeveloped roadless area" (39.2 percent), ranking below picnicking (54.7 percent) but above fishing (34.4 percent) and camping at a developed site (25.2 percent). Further down the list were driving off-road (24.0 percent), bicycling or mountain biking (16.2 percent), and hunting (14.2 percent).[23]

Asked about possible management objectives for these national forests, more than two-thirds felt that designating more wilderness areas was "important" to them (67.1 percent, with 41.4 percent feeling this was extremely important). Most objectives that rated higher with the public—including protecting old growth forests (85.3 percent) and trails for nonmotorized recreation (68.7 percent)—are enhanced by wilderness designation. By comparison, only half as many felt it was important to pave new roads (34.5 percent) and less than one-quarter felt it important to expand access for motorized off-road vehicles (22.8 percent).

Two-thirds of the public living within the market area for these national forests felt it important to designate more wilderness areas. This support was remarkably consistent among the subgroups identified in the poll—between long-term versus shorter-term residents of the region (64.6 percent of residents for more than twenty years favored more wilderness, 69.6 percent of those living in the region for less than ten years); between owners of rural land (66.5 percent) and nonowners (67.2 percent); and between white/non-Hispanic (66.2 percent) and non-white respondents (69.4 percent). While there was a slightly greater spread between working people (69.2 percent) and retirees, the latter also registered strong majority support (60.9 percent).[24]

U.S. Forest Service survey scientists offered their advice to the planners concerning the twelve issues the planners had identified at the outset, phrasing each as a management direction well supported by the those polled. Included was "Recommend additional roadless areas on National Forest System land for wilderness designation."[25] Noting that the growing, diversifying population of the region means shifting attitudes toward management of the forests, the authors conclude

> The trend has been toward more sensitivity for maintaining the natural condition and appearance of lands and forests. While we generally are a consumptive society, people express deep concern and caring about the future of natural resources and lands, such as National Forests. Overall, with some modest variations, there is agreement among people of different ages, races, employment status, places of residence, and education that careful management of National Forests to assure clean water, sustained healthy forests, wildlife habitat and naturalness are of highest priority. This is clear direction for setting management emphasis in [new national forest plans].[26]

Every one of these highest public priorities is not only compatible with wilderness but will best be assured by using the full statutory strength of wilderness designation to protect more of the national forests for permanent protection and enjoyment.

The evidence of diverse polling across the country leads to the overwhelming conclusion that, were the people of the entire nation invited to be "at the table" as part of federal agency planning, they would express their overwhelming support for the strongest possible protection of a great deal more wilderness.[27]

The process of designating additional wilderness areas will not be easy; it never has been. The politics of wilderness are complex and challenging, but the consistent message of reputable polling is that the people want more wilderness protected in this way.

Real wilderness, big wilderness—country big enough to have a beyond to it and an inside. With space enough to separate you from the buzz, bang, screech, ring, yammer, and roar of the 24-hour commercial you wish hard your life wouldn't be. Wilderness is a beautiful piece of the world. Where as you start up a trail and your nine-year-old Bob asks, "Is there civilization behind that ridge?" you can say no and share his "That's good!" feeling.

—David Brower,
"Wilderness-Conflict and Conscience,"
Wildlands in Our Civilization

Chapter 7

Wilderness Politics

Wilderness is only truly protected when it has been designated by Congress, subject to the legal protections of the Wilderness Act, and surrounded by a boundary only Congress can alter. The lesson of wilderness preservation history traced here is clear: however sincere the promises of protection in administrative orders and plans may be, anything less than statutory protection is temporary at best and illusory to boot.

More wilderness areas will be protected in the decades ahead, many more. In that work, grand visions and public enthusiasm, while necessary, are not in themselves enough. Most wilderness areas have been designated by Congress, and new ones will be only when local people who know and love a particular wild place develop their own proposal and organize to build broader local support among their neighbors. Understanding how more wilderness can and will be preserved is a matter of understanding the practical politics of legislative action.

The Power of Public Opinion for Wilderness

When a small group of wilderness leaders decided in the late 1940s to reorient the wilderness movement toward a wilderness law, choosing protection through the politics of legislation, they made a deliberate but risky choice—risky because there was no assurance they could succeed, and real perils if they tried but failed. Well aware of the risks, they nevertheless judged that statutory protection would bring stronger, more permanent protection, and that it could be a more successful avenue by which to extend protection to more lands. So, they went to work on the practical politics of making it so.

At the time there were respected contrary voices, and there were doubters still in 1964 when the law was signed. One participant observed that some conservation leaders felt at the outset that the first bill in 1956 "was at once too broad in coverage and too specific in detail—that it would arouse every conceivable type of opposition from every conceivable group for every conceivable reason."[1]

Four decades later, we can conclude that thanks to the foresight and perseverance shown by that generation of wilderness leaders and their congressional champions, the prospects for preserving wilderness in perpetuity took a huge turn for the better in 1964. As the campaign to enact the Wilderness Bill neared its conclusion, the Sierra Club's David Brower wrote that enacting the 1964 law "took time because the meaning of wilderness had not yet achieved the public understanding it now has—in large part because of the battle for the Wilderness

Bill."[2] In fact, as Harvey Broome observed, it was *because* of the national campaign for the Wilderness Bill that "the concept of wilderness became an imperative in American life."[3]

In taking their advocacy for wilderness out of the corridors of agencies in Washington, D.C., instead making their case to the public at large in order to influence Congress, advocates got a measure of the immense sympathy and support the American people feel for preserving a decent sampling of wilderness. It was a lesson that a generation of leaders—Brower, Zahniser, Brandborg, and their colleagues—learned in defending wilderness from threats in the Echo Park fight. They learned that the American people "get it" about wilderness and can be rallied to its support in overwhelming numbers.

By large majorities Americans favor protecting wilderness as strongly as possible, and protecting more than has yet been designated by Congress.

This broad public support is not the final word about the politics of wilderness, but it is the first and most fundamental political fact—not an opinion but a *fact* documented in scores of public opinion polls. As the *Daily Astorian* in Oregon editorialized in the spring of 2004, "in any given year, a small fraction of Americans crosses a wilderness trail head and visits these fragments of a bygone Eden. Yet they enjoy a wide though nearly invisible popularity."[4] This latent "people power" is the fuel for the politics of wilderness protection.

Wilderness Opponents

Some fervently wish this broad public support for preserving wilderness would fade. Generations of timber and oil company lobbyists, joined more recently by off-road vehicle groups and industry-financed right-wing think tanks, have eagerly predicted the decline of public support for protecting wilderness and wilderness wildlife.

As I was writing this, the Salt Lake City *Deseret Morning News* quoted Clark Collins, a founder of the BlueRibbon Coalition, saying many recreational users of the public lands "are falling out of love" with wilderness designation.[5] Of course, the only recreational users this alliance of off-road vehicle manufacturers, dealers, and devotees speaks for are those who choose to ride off-road vehicles—and only some of those. Most ORV enthusiasts use their vehicles responsibly; they have a legitimate claim to some portion of the public lands, to be sure. But perhaps many members of the BlueRibbon Coalition never fell *in* love with wilderness—the real thing—to begin with. After all, the founders of the wilderness concept defined wilderness first and foremost by exclusion of all things mechanical, protecting areas where, in Bob Marshall's phrase, people can escape from lives "saturated by machinery."[6] And that has been the legal definition for four decades.

Off-road vehicle lobbyists would like to convince the media and members of Congress that the debates over wilderness are just a face-off between two contending sorts of outdoor recreation, nothing more than a choice between hiking and backpacking (often with the implication that this is only for the young and the fit) versus "family recreation" with an off-road vehicle. To drive home their point, the BlueRibbon Coalition says that it is dedicated to "preserving our natural resources for the public instead of from the public."[7] But, of course, millions of Americans use and value wilderness areas precisely because these areas offer them the opportunity to get away from what Aldo Leopold called the "manifestations of gasoline,"[8] from what he and the other founders of The Wilderness Society termed "mechanical invasion in its various killing forms,"[9] from dependence on anything with wheels.

Antiwilderness lobbying by the BlueRibbon Coalition and similar groups does not represent all off-road vehicle enthusiasts or even a majority, any more than the Sierra Club or The Wilderness Society represent all wilderness lovers. Among off-road vehicle users surveyed by the Forest Service's National Survey on Recreation and the Environment, 73 percent strongly or somewhat favored protecting more wilderness in their own states. In the home region of the BlueRibbon Coalition (Idaho, Wyoming, Utah, Oregon, and Washington), 51 percent of off-road vehicle users favored more wilderness in their own states.[10]

For wilderness opponents, "access" has become the code word of choice, with the implication that wilderness areas are uninviting and unavailable to most people without vehicles and more roads:

- Opposing one wilderness proposal pending in Congress in 2004, the Washington State Farm Bureau claimed it "will permanently and unnecessarily restrict the public's use of and access to these lands." They add that "productive agriculture should not be sacrificed to wilderness encroachment"—as though they have hopes of plowing our national forests.[11]
- *Dirt Bike Magazine* rallies its readers to protect recreational access to public lands—as though barring motors and machinery equates to barring people or is some stealthy assault on the American way of life.
- A flyer distributed by opponents of a 2004 wilderness proposal repeatedly trumpeted the heading "*ACCESS DENIED*," asserting that "access will be denied to over 100,000 acres, previously used by many people involved in horse riding, mountain biking, snowmobile, motor cycling and four wheeling due to lack of roads. ... "[12] (They had their facts wrong; horse riding and packing are a common and respected form of wilderness use.)

Of course, the issue is not barring people from wilderness. As Representative John Saylor said, the recreational function of wilderness is about

> putting the American citizen back on his own two, God-given feet again,
> in touch with nature. Let's get him away for a blessed few moments from
> the omnipresent automobile, if we can, rather than catering to what is
> easiest and least exerting. And let's certainly not turn every quiet
> wilderness pathway into a "nature trail" for motors.[13]

Rather, the point of wilderness protection is to maintain a decent sampling of public lands free from all that would destroy its wilderness character. And what better way to reach those area-by-area choices in a democracy than by the votes of those elected to represent all of the people.

As it makes choices about proposed wilderness areas, Congress is the forum for contending viewpoints, very often strongly stated. And those who stand to benefit from minimizing further wilderness designations are, of course, drawn together. Thus, *Dirt Bike Magazine* urges it readers to "tie access, forest health, and the economy together" in writing their legislators.[14] The head of the Utah Shared Access Alliance asserts that "much public support for wilderness results from ignorance of its impact on public recreation. ... The real outcome of dropping millions of acres of land into a management black hole is the devastation of rural economies and the exclusion of the vast majority of family recreational visitors."[15] Off-road vehicle groups ally in a coalition of common interest with logging and oil and gas companies under the banner of access—access for your logging, access for her snowmobile and his dirt bike, access for this fellow's mining and that company's oil and gas drilling. The outcome they seek is maximum access. The areas dominated by the kind of "access" many wilderness opponents desire will be, in effect, "locked up" to other public uses, except for those who choose to picnic or camp in industrial landscapes.

Wilderness proponents face an inherent disadvantage in this process, for the burden falls on us to get Congress to pass laws designating new wilderness areas, while opponents are at the luxury of working to block us, focusing their efforts on our initiatives while they already have access for their uses on an enormous portion of public lands. It is a burden of proof we willingly accept in order to achieve the strongest possible statutory protection. Opponents have their own disadvantages, for theirs is an unholy alliance of diverse, even conflicting interests.

Rather than owning up publicly to their role and their behind-the-scenes lobbying, commodity developers often hide behind the "family recreation" face of off-road vehicle groups, but such deceptions have a way of coming to light. In one recent legislative campaign for wilderness, the chief timber lobbyist sat anonymously in the back of the Senate hearing room letting others, less obviously self-interested, be the public

[Americans can enjoy] a twofold civilization—the mechanized, comfortable, easy civilization of twentieth-century modernity, and the peaceful timelessness of the wilderness where vast forests germinate and flourish and die and rot and grow again without any relationship to the ambitions and interferences of man.

—Robert Marshall, "Maintenance of Wilderness Areas," in *Proceedings*, 30th convention of the International Association of Game, Fish and Conservation Commissioners, August 31–September 1, 1936 face of opposition. The same timber lobbyist wrote his letters to the editor without revealing his special interest affiliation. A new antiwilderness Web site showcased opposition from grassroots snowmobilers. This site—"Forests for People"—describes the organization as "a citizen grass roots group dedicated to public education regarding the importance of managing our natural resources without violating our Constitution nor jeopardizing our individual rights and freedoms."[16] The kicker: this Web site is legally registered to the same anonymous logging lobbyist, a fact nowhere disclosed. In adopting such stratagems, lobbyists for extractive industries and developers seek to emphasize the notion that there is broad public opposition to protecting more wilderness.

> In the background of man's history the years of the mechanical are insignificant compared with the years before machinery was known. Men are not yet adjusted to a world exclusively made of machinery. ...
>
> Our fast dwindling wilderness remnant requires that the burden of proof should be reversed, and only those roadless areas should be invaded where the urgency of developing transportation systems is unquestionable and the particular [unprotected] wilderness to be invaded can reasonably be spared. ...
>
> —Robert Marshall, "A Plea for the Old Wilderness," *New York Times Magazine*, April 25, 1937

What the Polls Say about Public Attitudes

When the Campaign for America's Wilderness, looking back at four years of polling, published a comprehensive analysis of all recent public opinion surveys about wilderness in January 2003, we could not find *any* professionally respectable polls showing majority opposition to protecting more wilderness.[17] But we found dozens of polls that showed strong public support—nationwide polls, state polls, even a poll at the county level in rural Oregon. Whether sponsored by newspapers or environmental groups, all were performed by reputable polling firms. *Every* poll showed strong support for protecting *more* wilderness areas. To allow readers to assess the fairness of each poll, our analysis included the exact wording of each question and the details of each survey sample.

These diverse, recent poll results show that public support for more wilderness enjoys majority support across age groups, political affiliations, regions of the country, and urban and rural areas. While the absolute size of public support for more wilderness varies among polls, the spread between support and opposition in these survey results is dramatic.

The most intensive polling about wilderness is done by the federal government itself. Managed by the research arm of the Forest Service, the National Survey on Recreation and the Environment has been repeated for decades and uses an unusually large sample, giving its findings great statistical reliability. From top to bottom, the NSRE findings probing attitudes about wilderness confirm those of the other polls.

- Asked about protecting more wilderness in their own states, 69.8 percent were in favor (42.5 percent strongly so); only 12.4 percent oppose (just 6.0 percent strongly).[18]
- Analyzing these results, on the same question rural residents living outside metropolitan areas (as defined by Census Bureau distinctions) overwhelming favored protecting more wilderness in their own states—62.8 percent (34.6 percent strongly)—with 18.8 percent opposing (10.4 percent strongly).
- This rural support was only a bit lower than the support among metropolitan area residents (71.5 percent).
- On the same question, 75.2 percent of Hispanic Americans (who were offered the survey in either Spanish or English) favored protecting more wilderness in their own states (47.6 percent strongly so), while just 6.8 percent opposed (3.2 percent strongly).[19]

The Forest Service polling did not ask these wilderness questions in a vacuum. Respondents were first informed of the tradeoffs involved:

> The Wilderness Act of 1964 allows Congress to preserve certain federal lands in their wild condition. These lands cannot be used for purposes such as timber harvesting, developing ski resorts, or highways. To date, Congress has added 625 areas to this National Wilderness Preservation System to protect wildlife, scenery, water, and recreation opportunities, and to keep these areas wild and natural.[20]

The Forest Service researchers, led by Dr. Ken Cordell, have repeatedly probed what it is about the values, uses, and benefits of wilderness that attracts such public support—the *why* behind the consistent majorities for preserving more wilderness areas. Having rigorously tested their categories, they presented respondents with a list of wilderness values, asking them to rate how important or unimportant each of these wilderness benefits was to them. While the American people assign very high importance to all of these values of wilderness by overwhelming margins, when their responses are ranked, one counts down to the ninth out of twelve before reaching on-the-ground recreational use within wilderness areas.

This explains why Congress understands that wilderness politics are not as simple-minded as choosing between two competing kinds of outdoor recreation. The American people feel very strongly that wilderness areas are important for multiple uses and benefits. In a comparative analysis, Dr. Cordell and his colleagues found that the degree of importance respondents assigned to each of these changed from the 1994 results, all shifting toward greater importance in 2000 survey results. Reviewing the findings of other researchers as well, they report

> There has been speculation that a fundamental shift has occurred in what people value in forests and other natural environments. This suspected shift

Percentage Spread between Wilderness Support and Opposition in Selected Public Opinion Polls

("Support"/"oppose" refer to designating additional wilderness areas)

Poll Details	Support (percent)	Oppose (percent)	Spread (points)
Nationwide poll, Zogby America, 2003, N=1,001 respondents (Congress has designated 4.7% of land in U.S. as wilderness; do you think this is too little or too much?)	64	6	58
Nationwide federal poll, National Survey on Recreation and the Environment (NSRE), 2000-2001, N=15,620 (Has Congress designated too little wilderness vs. too much wilderness?)	49	6	43
NSRE, 2000-2001 - all respondents, N=10,382 (Designate more wilderness in your own state; support or oppose?)	70	12	57
NSRE, 2000-2001 - Hispanic respondents, N=1,539 (Designate more wilderness in your own state; support or oppose?)	75	7	68
Nevada poll, Mason-Dixon Polling & Research, 2001, N=625 (Too little wilderness designated in Nevada vs. too much wilderness, noting off-road vehicles not allowed in wilderness areas?)	56	4	52
Vermont poll, University of Vermont Center for Rural Studies, 2002, N=472 (Should more wilderness be designated on Green Mountain National Forest?)	73	20	53
Idaho poll, Moore Information, 2004, N=500 (Should 500,000 acres of new wilderness be designated in SW Idaho's Owyhee County, noting off-road vehicles not allowed in wilderness areas?)	69	25	44

DATA SOURCES

May not add to 100% due to rounding. Full detail on each poll except the last, as well as many others, is found in *A Mandate to Protect America's Wilderness: A Comprehensive Review of Recent Public Opinion Research* (Washington, D.C.: Campaign for America's Wilderness, 2003), http://leaveitwild.org/reports/polling_report.pdf.

Benefits of Wilderness - Which Are Most Important to People?

Rank	Benefit of wilderness	Extremely / very / moderately important	Slightly / not at all important / don't know / refused
1	Protecting water quality	97.3%	2.7%
2	Protecting air quality	97.2%	2.8%
3	Protecting wildlife habitat	96.5%	3.5%
4	Legacy value: knowing future generations will have wilderness areas	94.7%	5.3%
5	Protecting rare and endangered species	94.0%	6.0%
6	Protecting unique wild plants and animals	93.1%	6.9%
7	Providing scenic beauty	92.6%	7.4%
8	Existence value: knowing that wilderness areas exist	91.7%	8.3%
9	Providing recreation opportunities	90.4%	9.6%
10	Option value: knowing that in the future I will have the option to visit a wilderness area	90.2%	9.8%
11	Preserving natural areas for scientific study	85.4%	14.6%
12	Providing spiritual inspiration	81.5%	18.5%

DATA SOURCE

National Survey on Recreation and the Environment poll (U.S. Forest Service Research), 2000-2001, N=5,239 respondents nationwide. Question: "Wilderness areas provide a variety of benefits for different people. For each benefit I read, please tell me whether it is extremely important, very important, moderately important, slightly important, or not important at all to you."

is away from the dominant social paradigm (that emphasizes economic growth and human dominance and use of nature) toward a new environmental paradigm (emphasizing sustainable development, harmony with nature, and a balance of human and nonhuman uses and nonuses).[21]

The politics of protecting wilderness rest on this bedrock support by strong majorities—not only nationally, but also in state level polls, and not only among metropolitan residents, but with rural dwellers as well. That younger voters and the rapidly growing constituency of Hispanic Americans register even stronger support for the benefits of wilderness and for protecting more of it than does the population as a whole suggests a politically important harbinger of consistent and even growing wilderness support in the future.

The Realities of Legislative Politics

My colleague Tim Mahoney, a master wilderness lobbyist, likes to explain the practical politics of wilderness preservation using the real-world parallel of the sweeping reform President Bill Clinton proposed for our much-troubled American health care system. Clinton originally proposed a complex, highly detailed reform plan, but, in the face of the political realities of Congress, that ambitious proposal collapsed on its own weight and complexity. A few years later, a *New York Times* story was headlined "One Step at a Time: Clinton Seeks More Help for Poor and Elderly." This 2000 story reported

> It seems more and more that ecological and existence values are central to American's viewpoint on wilderness. It is increasingly clear that protection of the lands within the [National Wilderness Preservation System] from development and exploitation is what most Americans want.
>
> —H. Ken Cordell, Michael A. Tarrant, and Gary T. Green, "Is the Public Viewpoint of Wilderness Shifting?" *International Journal of Wilderness*, August 2003

The budget unveiled today by President Clinton shows his extraordinary persistence as an architect of domestic policy. In health care and other social programs, he is pursuing many of the same goals as in 1993, though often by different means. ... Mr. Clinton described his approach in his State of the Union Message last month. "The lesson of our history, and the lesson of the last seven years," he said, "is that great goals are reached step by step, always building on our progress, always gaining ground."[22]

It is a reality of legislative bodies that they seldom and perhaps never enact sweeping visions. It is inherent in the give-and-take of legislative bodies whose members stand for reelection every few years that they can only reach decisions by doing what proves to be do-able. They work toward great goals in increments, in decisions powered by mutual accommodation and compromise. And that, in a nutshell, is how America's wilderness areas are being saved. Additional areas are designated by Congress year in and year out, incrementally, in response to grassroots advocacy and citizen support.

After the 2002 election in which Republicans strengthened their hold on the U.S. House of Representatives and took control of the Senate, the *New York Times* reported that "businesses and industries that donated millions of dollars to elect

Republicans are mapping out strategies to take advantage of the party's sweep in Washington."[23]

Despite this generalization, it is dead wrong to conclude that all Democrats support or all Republicans oppose wilderness legislation. Such an assumption flies in the face of a history of thorough bipartisan leadership for wilderness preservation in Congress, which in turn reflects broad public support. Of course, in our system interest groups do tend to line up with those individual politicians who are most often sympathetic to their causes, and some conservatives—in both political parties and in right-wing think tanks—have an antiwilderness agenda. When those hostile to wilderness preservation gain control of some or all gatekeeper roles in the congressional leadership (or in the executive branch), those who stand to benefit from developing roadless public lands are emboldened.

Not only do wilderness politics not fit neatly into a Republican versus Democrat analysis, it is also an error to assume that all or even most conservatives are wilderness opponents. After all, the words *conservative* and *conservation* derive from the same root. Conservative theorist Roger Scruton describes the conservative temperament in words that could apply as well to any wilderness advocate resisting further development of favorite wild public land:

> Conservatism is not an economic but a cultural outlook, and ... it would have no future if reduced merely to the philosophy of profit. Put bluntly, conservatism is not about profit but about loss; it survives and flourishes because people are in the habit of mourning their losses, and resolving to safeguard against them.[24]

As president, Republican Teddy Roosevelt was the critical player in shaping modern conservation ideas. After visiting the Grand Canyon in 1903, he exhorted a crowd

> I could not choose words that would convey or that could convey to any outsider what that canyon is. I want to ask you to do one thing in connection with it in your own interest and in the interest of the country—

This nation began as an "errand into the wilderness." The first task was to tame that wilderness, but as Wallace Stegner says, we need to preserve some of it "because it was the challenge against which our character as a people was formed." Most people will never care a fig about wilderness. Perhaps that means it is an elitist concern. But so what? ... Enjoyment of wilderness may not be spontaneous and "natural." It may be a learned process, inviting and even requiring reflection. But it is nonetheless valuable for being an aristocratic pleasure, democratically open to all.

—George F. Will,
Newsweek, August 16, 1982

to keep this great wonder of nature as it now is. (Applause.) ... I hope you
will not have a building of any kind, not a summer cottage, a hotel or
anything else to mar the wonderful grandeur, the sublimity, the loneliness
and beauty of the canyon. Leave it as it is. Man cannot improve on it; not a
bit. The ages have been at work on it and man can only mar it.[25]

These are not old-fashioned, cast-off ideas. A fundamental tension existed in
Roosevelt's day between those conservatives (of both parties) whose priority
allegiance was to corporate enterprises such as logging, mining, and oil and gas
development, and those with allegiance to the value of conserving all natural
resources, including preserving parks and wilderness areas. Even in the sharply
polarized politics of recent years, this same tension persists among conservatives.
One prominent Republican, Arizona Senator John McCain, addressed this tension in
a 1996 column in the *New York Times*, lamenting that so many Americans ask "have
Republicans abandoned their roots as the party of Theodore Roosevelt?" He noted
that "too often the public views Republicans as favoring big business at the expense
of the environment" and warned his party that

> Republicans should not allow the fringes of the party to set a
> radical agenda that no more represents the mainstream of Republicans
> than environmental extremists represent the mainstream of the
> Democratic Party.[26]

Senator McCain points out that there are also "hopelessly partisan environmen-
talists," a useful reminder that polarizing wilderness preservation issues along
partisan lines is, to put it bluntly, dumb political strategy. Despite a superficial
assumption of strong partisan differences, the reality is more often that many
members of Congress of both parties demonstrate broad nonpartisan support for
wilderness preservation. The real workshop of wilderness politics is not in the
rarified hothouse of Washington power politics but in the perceived public opinion
and politics of the individual states that shape the attitudes of members of each
congressional delegation. This is why the real work of wilderness protection lies in
organizing grassroots support for wilderness among the local citizens whose love of
their surrounding landscape has enormous, but often untapped political potential.

How does this play out in the real world of congressional wilderness
designations? For the most part, wilderness preservation decisions have been
thoroughly bipartisan. This can appear to break down at times when key committee
and leadership roles are held by anti-environmental members of Congress or when a
local congressional delegation is so staunchly antiwilderness that the more common
pattern of accommodation and compromise falls apart, as has been true in recent
years in Alaska and some other western states. But bipartisan wilderness designation
bills can make progress even then.

Wilderness and Compromise

So we come to a fundamental reality of wilderness politics in Congress: nothing in the Wilderness Act repealed the essential traits of legislators and the informal behavior of legislative bodies that lie behind the formal procedures by which they make laws. In our diverse and pluralistic democratic society, even those in the minority on a particular issue are heard, and compromise and mutual accommodation are part of the outcome. As a result most wilderness laws, beginning with the Wilderness Act itself, finally pass Congress by lopsided and bipartisan majorities—and often with unanimous approval.

Among advocates on both sides of wilderness-versus-development debates there is often an ideological disdain for compromise. This disdain—which I most emphatically do not share—is indulged in by those on both "sides" who are apparently so comfortable with the rightness of their views that they aspire not to persuade but to impose their ideas on others. And, after all, decisions about land uses such as wilderness designation are judgments of values, about which honest disagreement is possible. Few of us are willing to defer to the righteously emphatic assertions of those with whom we contend in political debates. Compromise can be altogether honorable, so long as one remains true to fundamentals.

For legislative bodies to function at all, mutual accommodation and compromise are the necessary stock-in-trade of legislators. As one experienced Senate staff member turned academic put it, most legislators "of necessity, acquire an intensive interest in learning the other man's viewpoint and develop thereby high proficiency in bargaining, yielding, and working out compromises."[27]

As a senator, Hubert H. Humphrey was a fighting Democrat, the very icon of unrepentant liberalism in his era. Humphrey was schooled in the realities of legislative compromise at the hands of the master, Senate Democratic leader Lyndon Johnson, who warned him that otherwise "you'll be ignored, and get nothing accomplished." As Johnson's biographer, Robert A. Caro, describes, using Humphrey's own words, Humphrey got the lesson:

> Not only, he was to say, was compromise not a dirty word; those who refuse to compromise are a threat; "the purveyors of the perfect," as he came to call them, "are dangerous when they ... move self-righteously to dominate. There are those who live by the strict rule that whatever they think right is necessarily right. They will compromise on nothing. ... These rigid minds, which arise on both the left and the right, leave no room for other points of view, for differing human needs. ... Pragmatism is the better method."[28]

Political scientist Bertram Gross notes that "Whether a compromise is wise or damned as opportunistic or whether a compromiser is accepted as a realist or accused as a traitor depends on where the observer stands and how much value he places upon the *quid* and the *quo*."[29] Across the political spectrum on Capitol Hill, members of Congress deplore "purveyors of the perfect," generally writing off those who, in their common parlance, "let the perfect be the enemy of the good." This unwritten reality of legislative behavior lying behind the parliamentary rules is the reason that the wilderness system grows by incremental decisions, rather than the enactment of some sweeping vision. Visionary though it was, the Wilderness Act was itself an incremental step, the product of compromise and accommodations that fueled its way to enactment. It did not, for example, extend to any of the hundreds of millions of acres administered by the Bureau of Land Management; that came later—but would not have come at all had there not been a Wilderness Act to begin with. As Senator Humphrey, champion of the Wilderness Bill from day one, said of compromise, it was "important that we inch along even if we couldn't gallop along, at least that we trot a little bit."[30]

When I met Senator Humphrey, introducing myself as a student of the history of the Wilderness Act, his immediate response was, "Oh, Zahnie." He spoke with abiding affection of his memories of working with Howard Zahniser to protect wilderness. Humphrey and Zahniser shared visions of social progress across many fields. Like Humphrey, Zahniser was devoted to a model of legislative advocacy in which those with differing views are respected. Zahniser was ever eager to persuade those who opposed the Wilderness Bill, believing he could work things out, and willing, while maintaining the fundamentals of his own position, to accommodate their interests as far as he could.

Zahniser was not immune to frustrations with the legislative process. In the midst of the eight-year campaign for the Wilderness Bill he wrote

> no one, I suppose, knows better than I do how wearying it is to continue over a period of years what is essentially the same program. I can well sympathize with all who grow tired of having to repeat essentially the same efforts Congress after Congress. I surely am eager to go on to the next step.[31]

To some zealous adherents on each side of an issue, the idea of compromise is anathema. But few of us would wish to live in a society where decisions are awarded to those who are most zealous, simply because they are. In achieving protection for as much wilderness as we have, virtually everything we have accomplished has come because members of Congress on all sides of the issues have respected the fact that wilderness advocates have a track record of trying to find accommodations with others, and respecting the merit of responsible compromise.

—John A. McComb,
longtime wilderness lobbyist and
former head of the Sierra Club's Washington, D.C., office

Zahniser felt that broadening the consensus for the Wilderness Bill would make the protection of wilderness just that much stronger. If the very different views of the House committee chairman, Representative Aspinall, could be accommodated, Zahniser wrote to a colleague, "our consensus will be the broader and our prospects the better" for the future of wilderness preservation.[32]

Just weeks before his death, Zahniser gave a reflective speech to a wilderness conference in Oregon. He proclaimed that through the previous seven-year effort, with all of its frustrations, including Aspinall's blocking of the Wilderness Bill a year and a half earlier, "rather than suffering the ills of cynicism, which are so prevalent in Washington, where so much is abstraction and so far removed from the real things ... I have felt that our Congress is in many ways a remarkable institution."[33] He extolled the virtues of the bicameral Congress, noting that dam builders had succeeded in passing the Echo Park dam authorization through the Senate three times, "but they didn't get it through the House" for that was where conservation groups and their congressional allies were able to mount their strongest opposition. Zahniser drew the lesson:

> That leads to the second outstanding characteristic that I have learned to emphasize in our Congressional government—in our whole government— that is this: it is very difficult for anybody in our form of government to get anything done that anybody doesn't want done. Now you can see right away, that's a pretty good characteristic of a large democratic government established by a people who have learned to fear tyranny and to fear over- government.[34]

There is a kind of legislative inertia built into the way Congress operates; of the thousands of bills introduced each year, only a relatively few are enacted. To pass, a bill must work its way through complex procedures in both the House and Senate. The process is a sequence of steps. At each, opponents can attempt to block a bill, often anonymously. To reach the president's desk and become law, identical versions of a bill must ultimately be approved by both houses. Reflecting on how he and his colleagues had blocked the Echo Park dam a year earlier, Zahniser emphasized that their adroit use of this inherent legislative inertia was the key. Now, said Zahniser in that 1963 talk, "we have been enduring a similar working-out in connection with the wilderness legislation."

> We are advocating a program for the people of the United States of America. In Congress assembled—by the Senate, by the House—we are asking the whole people to espouse something that we, in our conviction of the public interest, have come to regard so highly that we will put great effort into it. ... We are asking for a national consensus, and the significance may not be how far we can move with this important first step

but in the fact that so many people take that step. When the wilderness law is enacted it will be the whole nation who will be for it. ... So our process through these years has been one of widening the consensus to the point where it comprises the majority. In our form of government, with characteristics that I have reported to you, we don't force—we persuade—we try to meet objections.[35]

The Geography of Wilderness Politics

Successful wilderness proposals are most often bipartisan wilderness proposals, worked out in the give-and-take of public debate and legislative accommodation. But another political reality powerfully influences the politics of individual wilderness proposals. In an era of sharp political divisions across the spectrum of national issues, some general patterns have dominated our national politics in recent years. With only occasional shifts, majorities of voters in certain "red states" (or congressional districts) consistently send largely Republican delegations to Congress, while majorities in certain "blue states" (or districts) consistently elect Democrats. These patterns have shifted in the past and will evolve in the future, but they set the context in which current issues are considered.

More precisely, as regards wilderness issues the geopolitical pattern is not so much about political party affiliation as it is about relative alignment between each elected representative and development interests. In many places, people in rural congressional districts are more development oriented and often hostile to the federal agencies that administer public lands from which, of course, local people gain the greatest benefits. They elect officials who may reflect these opinions, and one could map this using selected votes on public lands and environmental issues to distinguish a different sort of "red" and "blue."

In assessing the politics of wilderness, the sharp red/blue distinction by party is a starting point. More important are the views of the senators from that state and the representatives whose districts are involved. In one very "red"—Republican—state, it is conservative Republicans who were leading efforts in 2004 to work out legislative proposals for major areas conservationists have long sought as statutory wilderness. As has happened elsewhere, legislation in these cases would likely include other provisions dealing with federal lands outside the wilderness, in that way helping broaden a consensus for legislation that accommodates many diverse interests—ranchers, off-road vehicle groups, and more. Whether the wilderness gains are "worth it" in view of such other provisions is part of the judgment to be made.

As implied here, a further aspect of the geopolitics of the American legislative process is the greatly disproportionate influence the local congressional delegation has on matters within their state. There is no point in arguing whether this is "right" or "wrong"; it is simple political reality. My colleague John McComb likes to emphasize the lesson enunciated by former House speaker Thomas P. "Tip" O'Neill in his famous axiom that "all politics is local." As John says, "success in achieving wilderness designations goes to those who understand and work with this reality."

Understandably, other legislators generally respect the prerogatives of their colleagues over issues in their own territory, just as they expect those legislators to respect theirs. In extraordinary circumstances, great national interest focuses on a proposal and the prerogatives of the local delegation are less definitive in decisions—the only cases of this have been with wilderness in Alaska. But there, too, Congress made many compromises and accommodations with the Alaska delegation. This is why ambitious bills sponsored by a member of Congress to protect wilderness in a state other than his or her own, while they can be valuable strategic tools for defensive purposes, have not been enacted in that form. The political prospects for a new wilderness designation improve greatly as grassroots organizing work builds the kind of local support that can move members of that state's congressional delegation toward embracing an acceptable proposal, or at least toward softening their opposition toward accommodation.

In some places during some periods, this emphasis on the influence of the local congressional delegation can impede progress of wilderness proposals, but those patterns tend to shift. And in many cases, the real need in order to break through with a stalled wilderness proposal is for still more grassroots organizing and persuasion. In all such places there is a strong degree of diverse but latent public support for protecting wild lands, for "keeping it like it is around here" and for protecting the very things for which statutory wilderness preservation is the proven best tool. For example:

- In Oregon, proposals to designate additional wilderness in the Mount Hood National Forest involved two House districts; one, represented by a liberal Democrat, includes a large part of metropolitan Portland, while the other, held by a conservative Republican, includes all of rural eastern Oregon. In rural Hood River County, an April 2004 poll by a Republican-oriented polling firm found that 57 percent of respondents supported the proposal to designate an additional 160,000 acres of wilderness (38 percent supported this strongly), versus 32 percent opposed. When reminded that dirt bikes and other motorized vehicles would be banned in the wilderness areas, just 25 percent of all respondents said this would make them less likely to support the wilderness designation proposal.[36]
- In Idaho, a 2004 poll asked voters about possible components of proposed legislation to protect portions of the Boulder-White Cloud Mountains. Told that the proposal would designate about 300,000 acres of wilderness, which would "remain open to most types of recreation, but new roads, mining and off-road vehicles would be prohibited," 67 percent voiced their support, versus 25 percent opposed. The pollster reported that

There is support for the Boulder-White Cloud Bill among all subgroups, including support among 68 percent of Democrats, 60 percent of Republicans and 53 percent of Independents. In addition, 63 percent of off-road vehicle users and 68 percent of snowmobilers support the bill. Among the most popular components of the bill is the designation of 300,000 acres of wilderness, which is supported by a 67 percent–25 percent margin. The wilderness component is supported by 85 percent of Democrats, 70 percent of Independents and 55 percent of Republican voters.[37]

Connecting Wilderness with People

The realities of legislative politics and political geography lead to an inexorable conclusion: protecting wilderness areas by law is a matter of building grassroots support and advocacy. This must be more than simply rallying the already persuaded. Wilderness organizers must identify, persuade, and enlist new constituents for wilderness, and reach a broad spectrum of people across lines of race, ethnicity, and political party alignments. In this work the central task is to connect people with wilderness goals at the level of fundamental community values.

I don't often hang out in cowboy bars in rural America. But I know that if one opened a conversation there about capital-w wilderness, one might be inviting a sharp conversation if not a fight. On the other hand, if the conversation came around to the basic idea that the nearby countryside is beautiful and wild, one would find broad agreement with the sentiment that it "sure would be nice if it could just stay like it is around here." In our advocacy, wilderness supporters firmly believe that the best tool for keeping wild federal land "like it is" is statutory wilderness. Persuading people of this is a matter of getting past preconceptions about "the W word" to the underlying value of sustaining cherished open spaces and the historic wild landscape and wildlife of the original America.

It is at this level of fundamental shared community values that wilderness support must be cultivated. What are those shared values? Here is one measure. In 1972 the people of Montana chose to elect delegates to a constitutional convention in order to rewrite the constitution of their state. Preliminary to all of the details in the new constitution they wrote, which the voters of Montana adopted, the preamble offers a stirring expression of these values: "We the people of Montana, grateful to God for the quiet beauty of our state, the grandeur of our mountains, the vastness of the rolling plains ..."

The Wilderness System of the Future

A look back at four decades of experience demonstrates that new wilderness areas can be protected. Cynics and zealots are found on all sides in these wilderness debates but do not have a track record of success, whereas grassroots organizing and legislative participation do. Zeal tempered by a genuine appreciation for the world in which rural Americans live and for the realities within which every member of

Congress acts—an approach so brilliantly modeled for us by Howard Zahniser—is what has built a diverse national system of wilderness areas in forty-four states. And the wilderness system will continue to grow in the same way:

- During the 106th Congress (1999–2000), President Bill Clinton signed laws designating more than 1 million acres of new wilderness in seven states.

- During the 107th Congress (2001–2002), President George W. Bush signed laws designating 526,000 acres of new wilderness in four states.

- Bills proposing new wilderness in many states were pending in Congress in the summer of 2004, among them proposals in which conservative Republicans were taking the lead.

- Still pending in Congress are presidential recommendations, some from decades ago, for wilderness designation on federal lands administered by each of the four wilderness agencies. A presidential recommendation has important stature before Congress, a status that does not lapse simply because action has not yet been taken.

Many current proposals are intended to expand existing higher elevation wilderness areas by adding lower elevation lands. These can enhance biological diversity, strengthening habitat protection for wilderness-dependent wildlife and promoting year-round wilderness-quality recreational opportunities. Often the addition of these lower elevation areas, where more development activities may have already occurred and more nonconforming uses may have already been established, also introduces more difficult land-use conflicts to resolve during congressional consideration.

How big should the National Wilderness Preservation System become? What are the dimensions of an ideal wilderness system? How much wilderness will be enough?

These questions have no answers—and can have none. The decisions to be made in the future lie in the future, not to be constrained today by Congress or anyone else. These lowercase-d democratic decisions are up to the people and to those whom they will elect to represent them.

One could hypothecate an ideal wilderness system, but such hypothecations would amount to little more than projections of the rapidly outdated assumptions and biases of whoever makes them. A group of conservation biologists might describe one ideal system, a group of equally eminent zoologists another, and a convention of ornithologists yet another. Hikers would have one vision, theologians another. The ideal wilderness system to the BlueRibbon Coalition will not look like any ideal at all to the board of directors of the Sierra Club. One lover of alpine meadows might conclude that we have preserved more than enough of the Mojave Desert ecosystem but not enough of his favorite terrain, while a devotee of vast Arctic wilderness areas might see little point in the far-smaller wilderness areas we have in New Jersey or Georgia—which are deeply treasured by their own advocates.

And what would be the point of such an exercise in describing some transitory ideal? After Pearl Harbor, would the Congress have felt constrained by some master plan for the ideal system of military training bases prepared in the 1920s? Did

Congress conclude, four decades after the designation of the first statutory national park (Yellowstone, 1872), that they had by then preserved all the desirable national parks and could declare the National Park System "complete"? Did they have a blueprint back then that included every new park unit established in subsequent decades? Would it make any greater sense to say that the wilderness system we have in 2004, forty years after the Wilderness Act, is complete?

In the final analysis, the answer is—and must be—left to the American people and the Congress they elect every two years. Would any of the diverse interests just mentioned really care to delegate their views of the question to any of the others? These are, after all, judgments about what is valued and what has priority to each of us among the many uses of the great common federal land estate owned by us all. Congress wisely decided in 1964 to reserve these choices to its exclusive decision on our behalf. In just the same way those lands now designated as statutory wilderness remain protected only because Congress has chosen, year after year, not to scale them back. We will protect more wilderness if and when Congresses in the future choose to add more lands.

Soon after the Wilderness Act became law, U. S. Supreme Court Justice William O. Douglas observed that "we look backward to a time when there was more wilderness than the people of America needed. Today we look forward (and only a matter of a few years) to a time when *all* the wilderness now existing will not be enough."[38]

The "right" answer to how much wilderness "should" be preserved is unknowable, but I will venture my own prediction: however much wilderness we Americans choose to designate and protect using the Wilderness Act, future generations are likely to judge not that we preserved too much, but that we preserved too little.

Chapter 8

The Challenges of Wilderness Stewardship

If "an enduring resource of wilderness" is to be preserved as the Wilderness Act commands, the requisite first step is that areas be designated by Congress and made subject to the management provisions of the law. Once that is done, however, it is not enough. Each area must then be administered, generation after generation, to assure that its wilderness values are indeed protected. As Vance G. Martin, president of the International Wilderness Leadership (WILD) Foundation has written, "if we do not manage [wilderness areas] wisely, their value will decline as surely as if they were never designated."[1]

"Guardians not Gardeners"

When talking about land, it is hard to avoid the word *manage*, but when the subject is wilderness, the better word is *stewardship*. This word choice reminds us that in wilderness areas the goal is that "the earth and its community of life" evolve and change subject only to *nonhuman* influences. At it heart—no, at its soul—the Wilderness Act is an act of humility toward the special portions of the natural landscape it protects. Howard Zahniser emphasized the "wilderness philosophy of protecting areas at their boundaries and trying to let natural forces operate within the wilderness untrammeled by man." He explained that the objective should be, as far as possible, to provide the protection needed to allow natural forces to operate with "the freedom of the wilderness."[2]

Zahniser got to the paradox, writing that "such tracts should be managed so as to be left unmanaged." He warned

> Some men see hills come together across streams of running water and then fancy dams there. Some have seen the apparent loveliness of deer and have fancied the elimination of "cruel" predators. Some men fancy timber management to tidy up the forest. All these see such areas as scenes for their own "skill, judgment, and ecologic sensitivity." They all threaten the still lingering possibility that some of the earth may be so respected as not to be dominated by man and his works.[3]

Dave Brower extolled the diversity of big sweeps of country, "diversity [that] is the result of land management by the Creator. The very best management."[4] And Zahniser captured the essence of wilderness stewardship in what is for me a favorite

125

phrase: "with regard to areas of wilderness we should be guardians not gardeners."[5] Parsing the congressional direction in the 1964 law, a federal judge echoed Zahniser's analogy: "Nature may not always be as beautiful as a garden but producing gardens is not the aim of the Wilderness Act."[6]

I emphasize this fundamental philosophy of wilderness stewardship because it provides the essential foundation for understanding the many and complex challenges faced by the administering agencies and their dedicated wilderness stewards.

The Wilderness Act's Stewardship Mandate

As the wilderness system has grown, it was inevitable that each of the four disparate agencies administering wilderness areas would develop its own unique interpretations and policies about stewardship issues. However, there is just one Wilderness Act and wilderness system. Regardless of agency jurisdiction, the guiding vision for administration should be that the National Wilderness Preservation System be administered with maximum consistency. For one thing, there are numerous places where a single wilderness expanse is comprised of contiguous areas administered by two or three agencies, with each agency's portion administered differently in various details. Some of these differences are intentional—hunting is allowed in a national forest wilderness area but not in the national park wilderness area to which it is contiguous. Regardless of these underlying differences, the act gives an identical, overarching stewardship mandate to each agency. Subsection 4(b) commands that

> Except as otherwise provided in this Act, each agency administering any area designated as wilderness shall be responsible for preserving the wilderness character of the area and shall so administer such area for such other purposes for which it may have been established as also to preserve its wilderness character.

Preservation of wilderness character thus supplements other agency purposes, constraining uses and actions that would conflict with the wilderness preservation mandate.

But what is the essence of the wilderness character the agencies "shall" protect? Where in the act do administrators look to find the meaning of the goal of their stewardship?

The framers of the Wilderness Act intended the first sentence of the two-sentence definition of wilderness in subsection 2(c) to establish the meaning of "wilderness character":

> A wilderness, in contrast with those areas where man and his works dominate the landscape, is hereby recognized as an area where the earth and

its community of life are untrammeled by man, where man himself is a visitor who does not remain.[7]

These words animate the act's wilderness concept. Without this definition, the subsection 4(b) mandate to preserve "the wilderness character of the area" would be cast adrift, left floating, tied to no clear and practical meaning on which administrators can base stewardship decisions.

At the heart of this goal for wilderness stewardship is the word *untrammeled*. No other word in the Wilderness Act is as misunderstood, both as to its meaning and its function in the law. At the command of the Wilderness Act, we preserve wilderness character by leaving "the earth and its community of life untrammeled by man." The Oxford English Dictionary traces *trammel* to Latin and eleventh-century Old French roots, meaning a kind of net used to catch fish or birds. Current dictionary descriptions of the word *untrammeled* include "unrestrained," "unrestricted," "unimpeded," "unencumbered," "unconfined," and "unlimited."

The definition of *wilderness* in the law contains two sentences (see below). The law was structured this way deliberately. Commenting on the two-part structure of the definition during the final Senate hearing in 1963, Zahniser noted that

In this definition the first sentence is definitive of the meaning of the concept of wilderness, its essence, its essential nature—a definition that makes plain the character of lands with which the bill deals, the ideal. The second sentence is descriptive of the areas to which this definition applies—a listing of the specifications of wilderness areas; it sets forth the distinguishing features of areas that have the character of wilderness.

The Two Definitions of Wilderness

(c) A wilderness, in contrast with those areas where man and his own works dominate the landscape, is hereby recognized as an area where the earth and its community of life are untrammeled by man, where man himself is a visitor who does not remain. An area of wilderness is further defined to mean in this Act an area of undeveloped Federal land retaining its primeval character and influence, without *permanent* improvements or human habitation, which is protected and managed so as to preserve its natural conditions and which (1) *generally appears* to have been *affected primarily* by the forces of nature, with the imprint of man's work *substantially unnoticeable*; (2) has outstanding opportunities for solitude *or* a primitive and unconfined type of recreation; (3) has at least five thousand acres of land or is of sufficient size as to make practicable its preservation and use in an unimpaired condition; and (4) *may* also contain ecological, geological, or other features of scientific, educational, scenic, or historical value. [16 U.S.C. 1131(c)](emphasis added)

The first sentence defines the character of wilderness, the second describes the characteristics of an area of wilderness.[8]

We need not rely solely on Zahniser's expression of intent, for in the legislative history, the law's sponsors repeatedly emphasized the two distinct functions of the two sentences in the definition. In 1961 Senator Clinton P. Anderson, chairman of the Senate committee, was the lead sponsor of the Wilderness Bill. Opening hearings that year, he explained his intent in the wording of his bill. As part of his detailed section-by-section explanation, he stressed that the bill "contains two definitions of wilderness":

> The first sentence is a definition of pure wilderness areas, where "the earth and its community of life are untrammeled by man. ... " It states the ideal.
> The second sentence defines the meaning or nature of an area of wilderness as used in the proposed act: A substantial area retaining its primeval character, without permanent improvements, which is to be protected and managed so man's works are "substantially unnoticeable."
> The second of these definitions of the term, giving the meaning used in the act, is somewhat less "severe" or "pure" than the first.[9]

In fact, over its eight-year legislative odyssey, every one of the lead sponsors of the Wilderness Act explicitly expressed the intent that the first sentence of subsection 2(c) expresses the "abstract" (Senator Humphrey) or "ideal" (Senator Anderson, Representative Saylor), distinct from the "more concrete details in defining areas of wilderness" (Senator James Murray, Democrat of Montana, who preceded Anderson as the committee chairman and lead sponsor), which are spelled out in the second sentence. What was the point of them all taking such care to spell out this distinction? And if the second sentence provides the "less 'severe' or 'pure'" definition that Congress uses to judge areas proposed for designation, what is the function of the first sentence, which "states the ideal"?

Given the precise word choices and the care taken in structuring the two-sentence definition in the Wilderness Act, it is beyond dispute that

- Designation questions of whether a newly proposed area of land meets the definition of wilderness in the act are judged according to the second sentence and is not about whether that land is "untrammeled." The word *untrammeled*, which applies once an area is designated, appears only in the "pure," "ideal" definition that serves a quite different function in the act.
- In reaching its designation judgements, Congress has applied the clear qualifiers— an area that "*generally* appears to have been affected *primarily* by the forces of nature, with the imprint of man's work *substantially* unnoticeable." As congressional committees and leaders have pointed out repeatedly over four decades, this language provides the practical criteria for entry of areas into the

National Wilderness Preservation System. This is also how a federal judge read this provision of the act, in one of the few cases where these issues arose.[10]

- The ideal definition in the first sentence has an equally important but different function. It is not mere congressional poetry, for the canons of statutory interpretation forbid such an interpretation. The function of this sentence—with its careful use of the word *untrammeled*—is to define the "ideal" (Senator Anderson), the "essence" (Howard Zahniser) of the wilderness character it is the duty of agency personnel to protect. It is, in fact, the definition of that pivotal phrase *wilderness character*. That is why this exercise in close reading of the two definitions is so important.

There is a supreme logic to this careful structure of the two definitions. Applying the practical criteria of the second sentence in subsection 2(c), the 1964 act itself designated numerous areas with a fading history of the "imprint of man's work," and many others have been designated in subsequent acts of Congress.

But however less-than-pure an area may have been when designated, once it is designated, the command of the act is to preserve the "wilderness character" of the area, restraining man's influences in order that the earth and its community of life are untrammeled by man. Each wilderness area is administered, in other words, to come closer and closer to the ideal of being the product of the forces of nature, its community of life untrammeled by man. But this has not always been well understood.

The Nondegradation Principle

As the Wilderness Act was initially being applied to designate new areas, the Forest Service took the position that "the criteria used in determining suitability or nonsuitability for inclusion of lands in the system is also our guide for administering areas once they are included."[11] But this was a misreading of the law and its legislative history.

This now-abandoned agency point of view reflects a not uncommon misunderstanding. Some assume that if Congress includes some evidence of impacts and disturbances of human activities (such as logging scars and roads) in a wilderness area, that means that the same kinds of human activities may therefore be allowed to invade other, wilder wilderness lands already designated. Of course, such incompatible activities are expressly prohibited by the Wilderness Act. Yet some nevertheless worry that allowing less-than-pure lands to be designated will be a precedent for the administrators to allow new development to invade other wilderness areas. But precedents don't work like that.

That a madman once got into the Vatican and took a mallet and chisel to Michelangelo's *Pieta* was a tragedy for art. It took the best restorers years to do what they could to repair the damage to this priceless sculpture (and it is no longer, to be sure, the pristine sculpture it once was). But that whole episode was certainly not a "precedent" establishing that other madmen with chisels would be welcome in the Vatican.

This is the essence of the nondegradation principle that lies at the very heart of the Wilderness Act.

A parallel nondegradation principle is found in the Clean Air Act, which does not allow polluting clearer air down to some lower quality set in minimum, health-based air quality standards. Rather, that act gives the government power to enforce pollution controls (even on distant polluting sources) in order to keep already pristine air quality in certain wilderness areas and national parks from being degraded.

In the same way, Congress carefully distinguished the stewardship ideal of wilderness from the practical criteria by which it chooses to designate areas. The nondegradation principle inherent in the structure and drafting of the law requires that whatever the level of ecological "purity" of an area when it is designated, it is to be administered thence forward toward the wilderness ideal. Senator Church summarized this intent of his fellow framers of the Wilderness Act, terming it "one of the great promises of the Wilderness Act [that] we can dedicate formerly abused areas where the primeval scene can be restored by natural forces.[12]

By expressing the ideal of "pure wilderness," its "essential nature," in the ringing first sentence of subsection 2(c), Zahniser breathed ecological life into the act's stewardship mandate to preserve "wilderness character." He and Congress thus set the goal toward which our stewardship of wilderness areas is to strive: whatever past human impacts there may have been, our goal is to free nature within these special places, as best we can, from the trammels and fetters of man's influence, so that wilderness may be—through our own humble self-restraint—areas "where the earth and its community of life are untrammeled by man."

Dealing with Nonconforming Uses

There are often nonconforming uses within a wilderness area—uses that were there at the time that area was designated and that are not consistent with the ideal of wilderness protection. The founders of the wilderness movement, and ultimately the Congress, understood that such nonconforming uses were simply the on-the-ground reality and accommodations would have to be made. The nature of those accommodations—allowance for continuing established grazing is one prominent example—is beyond the scope of this book.[13]

In his most important wilderness article, Bob Marshall noted that "certain infringements on the concept of an unsullied wilderness will be unavoidable in almost all instances. ... but ... the basic primitive quality still exists: dependence on personal effort for survival."[14] In 1949 Howard Zahniser expanded on this point, explaining the philosophy with which he later approached the preparation of the Wilderness Bill:

Where considerations of expediency or recognition of existing practices have permitted inconsistent wilderness use—such, for example, as domestic stock grazing within designated wilderness areas in the national forest system—such uses should be recognized as nonconforming and looked upon as subject to termination as soon as this can be done and done equitably for those immediately concerned. Such nonconforming uses should be permitted only when their temporary sufferance appears to be a means of insuring future values of the area."[15]

Consistent with the intention of the drafters and lead sponsors, protection of established but nonconforming uses and of private rights was emphasized by Senator Frank Church as he managed the Senate floor debate leading to passage of the Wilderness Bill.

"I want to note that, with this bill, it is possible to create a wilderness system without adversely affecting anyone. ... In these areas lumbering is already prohibited. Such grazing as presently exists may continue as before. It is not affected by the bill. Insofar as mining is concerned, in all the area covered by the bill there are only six mines in operation today, and those mines would continue in business, because the bill expressly provides that any restrictions that may apply in a wilderness area are made subject to existing rights.

"So we can pass the bill without adversely affecting anyone's rights, if we act now. That is how the committee approached its task. It was with this objective that the committee drafted the legislation so carefully and so cautiously. That is why we held such extensive hearings."[16]

Wilderness Stewardship in Practice

The challenges of wilderness stewardship are now the daily work of thousands of federal employees, aided by thousands of volunteers. One part of my work in recent years has been to speak about wilderness history at training sessions for federal wilderness stewards.

This wilderness stewardship training program was the inspiration of Representative Bruce Vento, a Minnesota Democrat, who proposed two new institutions, one to promote research to assist wilderness administrators and one to train agency personnel in the specialized concepts and skills necessary for wilderness stewards. In 1993 senior officials took up his idea, establishing the Aldo Leopold Wilderness Research Institute and the Arthur Carhart National Wilderness Training Center, both based in Missoula, Montana.[17] The dedicated expert staff of the Carhart Center is drawn from all four agencies, as an important step in working for integrated approaches based on the common factor for all wilderness areas, which is, of course, the Wilderness Act itself. As the founding director of the Carhart Center,

Connie Myers, puts it

There is one National Wilderness Preservation System that happens to be
managed by four federal agencies. We must remember and appreciate that
while each agency has its own unique culture, bureaucracy, and
interpretations, the Wilderness Act calls them to manage wilderness as a
system. The role the Carhart Center plays is to help lend consistency to the
interpretation and application of the Wilderness Act across four agencies
through training, information, and education. Only by working together
across agency boundaries can we hope to pass wilderness to the next
generation in better shape than when we inherited responsibility for its
stewardship.

When I participate in the Carhart training programs, I always try to stay for
several days in order to learn from the attendees—and these sessions always mix staff
from all four agencies—about the realities of their work on the ground. It is
heartening to listen to these dedicated stewards discuss the conundrums of
protecting the naturalness of wilderness ecosystems and the wildness of wilderness
recreational experiences. They revere the Wilderness Act as the touchstone for each
decision.

The key guide to the details of wilderness stewardship is found in subsection
4(c) of the Wilderness Act, captioned "Prohibition of Certain Uses." This
congressional direction applies both to public use and agency activities within
wilderness areas. It distinguishes uses that are flatly prohibited—commercial
enterprises and permanent roads—from agency activities that can be allowed under
very limited circumstances. This agency discretion is limited in one of the most
delicate of Zahniser's phrasings:

> except as necessary to meet minimum requirements for the administration
> of the area for the purpose of this Act (including measures required in
> emergencies involving the health and safety of persons within the area),
> there shall be no temporary road, no use of motor vehicles, motorized

---

To administer Isle Royale as a wilderness area, it is important to secure a personnel
which has a feeling for wilderness and an understanding of wilderness values; otherwise
the desire to be doing something to the areas will be hard to curb. The administrators
should be told that their success and achievements will be measured, not by projects
accomplished, but by projects sidetracked. In the management of a wilderness area, we
must somehow depart from the 20th century tempo of activity. ...

—Adolphe Murie,
Report on field study of potential Isle Royale National Park, early 1930s

equipment or motorboats, no landing of aircraft, no other form of mechanical transport, and no structure or installation within any such area.

The elegance of this language lies in its two-part test for allowing exceptions for agency activities—activities that are prohibited for public users. The test asks first whether the proposed exceptional activity is *necessary* and, second, if so, whether the particular means proposed to accomplish that necessary activity is the *minimum* to achieve that purpose. The Carhart Center distributes a detailed "Minimum Requirement Decision Guide" that equips agency officials with worksheets to document their decisions, case by case, on each request for such activities—use of chainsaws in unusual cases of mass blowdown of trees across trails, replacement of bridges, proposals for wildlife management installations, and such. Senator Frank Church, a leader in passing the Wilderness Act, explained the rationale for the minimum requirement decision process:

> We intend to permit the managing agencies a reasonable and necessary latitude in such activities within wilderness where the purpose is to protect the wilderness, its resources and the public visitors within the area—all of which are consistent with "the purpose of the Act." ... The issue is not whether necessary management facilities and activities are prohibited; they are not—the test is whether they are in fact necessary.[18]

In 1999 the wilderness coordinators in the Washington, D.C., headquarters of the four agencies asked the Pinchot Institute of Conservation to convene a panel of nongovernment experts to review the first thirty-five years of wilderness stewardship and develop recommendations. The resulting Wilderness Stewardship Panel was chaired by Dr. Perry J. Brown, dean of the School of Forestry of the University of Montana. In its September 2001 report, the Brown panel found "the National Wilderness Preservation System is more important to the American people than ever before."[19]

The expert panel found a continuing and priority need "to forge an integrated and collaborative system across the four wilderness management agencies." Ensuring a wilderness system for the future, the panel said,

> will require building an integrated, collaborative system across the two departments [the Department of Agriculture and the Department of the Interior] and four wilderness agencies. To manage the wilderness as a system means that each area is part of a whole, no matter which agency administers it. It means that all wilderness [areas] are subject to a common set of guidelines, and thus requires that such guidelines be developed.[20]

The Brown panel proposed eight principles for wilderness stewardship:

- Adhering to the Wilderness Act is a fundamental principle for wilderness stewardship.
- Preservation of "wilderness character" is a guiding idea of the Wilderness Act.
- United States wilderness is to be treated as a system.
- Wilderness areas are special places and are to be treated as special.
- Stewardship should be science-informed, logically planned, and publicly transparent.
- Nondegradation of wilderness should guide stewardship.
- Recognizing the wild in wilderness distinguishes wilderness from most other land classes.
- Accountability is basic to sound stewardship.

To "increase the probability that the Wilderness System will be sustained," the Brown panel recommended acceleration of developing management plans for each wilderness area to help guide stewardship and assure evaluation and accountability. The panel emphasized that the ultimate responsibility lies at a political level, with the secretaries of agriculture and the interior, and called upon them to "devise an organizational structure to make stewardship happen across the agencies."

Seeing Wilderness Areas in a Larger Context

Deepening ecological sophistication brings even greater challenges in stewardship of wilderness. For example, the issue of invasive nonnative species poses conundrums regarding how far, if at all, wilderness stewards should go in intervening to manipulate the environment toward the goal of restoring what we perceive to be more natural conditions. Yet, in a world warming under the impact of global climate change, human-caused ecological change occurs on so thorough a basis that sorting human-caused change from natural ecological progression may prove impossible.

The very fact that our ecological sophistication is growing could tempt us, at any given moment, to conclude that we have enough wisdom to know what well-intentioned manipulation is best—a temptation to ecological hubris that conflicts with the ecological humility bound up in the very idea of wilderness which is, in the final analysis, a human construct.

Ideals are essential, yet the dilemma is that wilderness areas are not islands inherently protected from all that goes on outside their congressionally protected boundaries. If the overall ecosystems and landscapes of our public land are to be properly managed, the lands outside of wilderness areas must be given appropriate protection. For example, lands adjacent to wilderness areas are very often vital corridors for wildlife species moving across their larger territory. Protection of wilderness habitats is important to those species, but so is protective management of these corridors lying *outside* of Wilderness Act protection.

Wilderness areas also protect wild scenic vistas—scenery large numbers of people enjoy from nearby roads, developed recreation sites, or their own nearby homes. The wilderness beyond adds an ineffable but invaluable quality to their experiences.

At the behest of logging companies that feared proliferation of buffer zones around wilderness that could limit their activities, Congress has routinely included in wilderness designation bills a provision prohibiting any buffer zones. An example is this language from a 1984 law designating areas in Washington State:

> Congress does not intend that designation of wilderness areas in the State of Washington lead to the creation of protective perimeters or buffer zones around each wilderness areas. The fact that nonwilderness activities or uses can be seen or heard from areas within a wilderness shall not, of itself, preclude such activities or uses up to the boundary of the wilderness area.[21]

If there cannot be buffer zones outside a wilderness area *because* it is a wilderness, then logging, second homes, or other development can creep right up to a wilderness boundary. If the goal is to avoid this, the inescapable implication of forbidding exterior buffer zones is that wilderness boundaries be fixed at the *present* edge of roads and development. In fact, Congress has commonly set wilderness boundaries on this edge-of-development basis. Congress has also expanded some agency wilderness proposals to include more of the sensitive lower-elevation lands, to better protect the full range of biological diversity and to better buffer higher peaks and ridges. For example, a congressional committee explained why it extended the boundary of the High Uinta Wilderness in Utah:

> The Forest Service's recommended wilderness boundary on the north slope of the range was adjusted to incorporate lower elevation lands in key areas. ... According to Forest Service and other wildlife data these additions comprise some of the finest elk, moose, and other wildlife habitat in the entire Uinta range. They also include some of the most popular horseback and hiking access trails into the wilderness and will further act as a buffer to keep vehicles and other development removed from the fragile high country that forms the core of the Uinta Range.[22]

The complexities and challenges facing wilderness stewards are both daunting and subtle. As one measure, the definitive text on the subject—*Wilderness Management: Stewardship and Protection of Resources and Values,* now in its third edition—runs to 640 pages.[23]

Just as people save wilderness through the designation process, so it is people who provide the stewardship for those wilderness areas Congress has chosen to protect. While problems and controversies do arise and there can be sharp differences between wilderness advocacy groups and some administrative decisions, the bottom line is that we are blessed by the high quality and dedication of the people who have chosen to dedicate their careers to this work. Connie Myers says

> The women and men drawn to work in wilderness bring a remarkable diversity of knowledge and skills. They include professionals trained in forestry; wildlife, fisheries, and plant biology; archeology; recreation; planning; fire; and ecology. But rather than viewing the landscape solely through the lens of their specialty, wilderness stewards have the uncanny ability to see the big picture—to appreciate that the whole of wilderness is greater than the sum of its parts, that the whole is what must be managed. These professionals share a remarkable, almost visceral passion for their work and have committed their careers to doing everything they can to ensure that wilderness endures for this and future generations.

"Inholdings" and The Wilderness Land Trust

As it passed the Wilderness Act, Congress recognized that some private and state-owned tracts existed within the boundaries of federal wilderness areas (and within areas added to the system since then). Many of these private "inholdings" originated as old homesteads, railroad grant lands, and mining claims under the archaic Mining Law of 1872. Unless and until acquired by the federal government, these lands are not subject to the protective provisions of the Wilderness Act.

Section 5 of the Wilderness Act assures that owners of such inholdings "shall be given such rights as may be necessary to assure adequate access." These nonfederal inholdings are in some cases the site of current conflicts or have the potential for conflicts. As one chief of the Forest Service wrote, "A wilderness area is vulnerable as long as it is shot through with private inholdings. ... The owners cannot be prevented from doing things on or with their lands, if they choose, that might be detrimental to wilderness values."[24]

The agencies work to acquire these lands by equal value land exchanges or by negotiating willing-seller purchases. Often, however, the opportunity to strike such deals moves faster than congressional appropriations and agency approval processes. Into this gap comes The Wilderness Land Trust, a nonprofit organization dedicated solely to opportunistically acquiring such lands, which it sells in turn to the federal agencies as funding and approvals allow. Upon federal acquisition, these lands are automatically subject to the same protective provisions as the surrounding federal lands.

While some acquisitions in wilderness areas are pursued by local land trusts or national groups such as the Trust for Public Lands and The Nature Conservancy, The Wilderness Land Trust was organized in 1992 to specialize in wilderness inholding issues. In its first decade the trust acquired and transferred more than 15,000 acres of inholdings in thirty-eight wilderness areas in seven states. The group—whose board I joined in 2002—now operates across the country and the mission has broadened.

While the trust takes no role in the politics of wilderness designation, it is stepping in to acquire inholdings in proposed as well as established wilderness areas. This can help remove side issues that complicate wilderness designation proposals. The trust has gained credibility with land owners for its appreciation of private-property rights, its fair dealings, and its creativity in tailoring innovative win-win solutions pleasing to land owners and the government.

Chapter 9

Wilderness in Other Realms

The federal Wilderness Act was, like the establishment of the first national park, a new idea. In the debate over passing the legislation, the concept of statutory protection for wild lands gained broad exposure. Not surprisingly, the result has been the spread of the wilderness-by-law idea to tribal lands, state lands, and lands far beyond U.S. borders.

Tribal Wilderness

Most American Indian tribes are sovereign nations whose relationships to the United States are defined by treaty on a government-to-government basis. Tribes have sovereign authority over their tribal lands. Many, perhaps even most tribes have off-reservation rights to fish and hunt and gather natural resources. These rights as well as tribes' deep interest in sacred sites lead them to want to protect the integrity of natural ecosystems on federal lands. In many places tribal lands are adjacent to existing and proposed wilderness areas, and Indians may make traditional use of those federal lands for religious and cultural purposes.

In many cases tribes are a key part of the coalition of support for protecting wilderness. In one recent example, the Cheyenne River Sioux Tribe called on Congress, by formal resolution of the tribal council in June 2003, to designate portions of national grasslands (administered by the Forest Service) in the Cheyenne River valley of South Dakota as wilderness. The resolution stresses that "grasslands are of vital historical and cultural importance to the Lakota People and the National Grasslands System is made up of lands acquired by the United States Government after its occupation of treaty lands of the Great Sioux Nation."[1] Noting that "the lands of the Cheyenne River Valley are Treaty lands that deserve the utmost respect and protection the United States Government can bestow," the tribe called wilderness designation "a necessary step toward protection of the lands that are essential to the cultural continuity of the tribal peoples of the Great Plains":

> the dwindling prairie ecosystem is in grave danger of extinction, the Four-legged; the buffalo; the coyote; the deer; the antelope; the badger; the prairie dog; the Black-footed ferret. The Winged Ones; the bald eagle; the golden eagle; the hawk; the owl; the plover depend upon the habitat of prairie to survive.

Tribes have also complemented federal wilderness designation with similar statutory protection for their own lands under tribal law. An outstanding example is the Mission Mountains in western Montana. A federal wilderness was designated by Congress in 1975 comprising 74,000 acres of national forest lands on the east slope of the mountain range. Lands on the west slope are on the Flathead Reservation. While earlier efforts had been made to protect wilderness values on these tribal lands, including by Bob Marshall when he was an official of the Office of Indian Affairs, they were opposed by the tribe, which had not been asked for its input or consent. In the 1970s tribal members circulated a petition asking the tribal council to stop timber sales proposed in the area and to consider some wilderness-like protection. Tribal member Thurman Trosper (a retired Forest Service official and former president of The Wilderness Society) suggested that the tribe have a consultant prepare a boundary and management proposal. In 1982 the tribal council of the Confederated Salish and Kootenai Tribes of the Flathead Reservation approved Tribal Ordinance 79A, which provided in part that

> Wilderness has played a paramount role in shaping the character of the people and the culture of the Salish and Kootenai Tribes; it is the essence of traditional Indian religion and has served the Indian people of these Tribes as a place to hunt, as a place to gather medicinal herbs and roots, as a vision-seeking ground, as a sanctuary, and in countless other ways for thousands of years. Because maintaining an enduring resource of wilderness is vitally important to the people of the Confederated Salish and Kootenai Tribes and the perpetuation of their culture, there is hereby established a Mission Mountains Tribal Wilderness Area and this Area, described herein, shall be administered to protect and preserve wilderness values.[2]

The resolution adopted the Wilderness Act's definition and approved a detailed management plan and user regulations for the 90,000-acre tribal wilderness. In a fine example of collaboration, the Forest Service has published a map that depicts both halves of this protected wilderness expanse and includes both the federal and tribal user regulations.

State Wilderness

In the 1970s a number of states established policies to protect wilderness on state-owned lands. Only some of these policies are based on state wilderness statutes, and the level of protection of the areas may be either statutory or administrative—a key distinction. Nonetheless, these are important steps to protect wilderness and to complement the federal system. As Dr. Chad P. Dawson and Pauline Thorndike wrote in a review of state wilderness policies in the *International Journal of Wilderness*:

State-designated wilderness areas add to the geographic and ecological diversity of areas given wilderness protection in the United States. In particular, Midwestern and Eastern states with limited federal lands can extend wilderness protection and stewardship to state lands and offer primitive recreation opportunities that might otherwise not be available.[3]

Some examples:

Model State Wilderness System: California

The California legislature enacted the California Wilderness Act in 1974, establishing "a California wilderness preservation system to be composed of state-owned areas designated by the legislature as 'wilderness areas' and units of the state park system classified as 'state wildernesses' by the State Park and Recreation Commission. ... "[4] The statute, closely modeled on the federal law, immediately designated the 87,000-acre Santa Rosa Mountains Wilderness Area and the 10,000-acre Mount San Jacinto State Wilderness. As of 2002 the California wilderness system comprised 466,000 acres in ten areas.

The state's San Jacinto area exemplifies one important role for state wilderness protection. It comprises most of the area in a state park contiguous to the 32,000-acre San Jacinto Wilderness. This federal wilderness area is in two units, lying north and south of the state park, with the state-designated wilderness filling the gap—in essence creating a 42,000-acre *federal-state* wilderness unit, all protected by statutory law.

State Wilderness in New York: "Forever Wild"

The protection of wilderness on state lands in New York played a key role in the history of America's wilderness protection policies. After the establishment of the state's Forest Preserve in 1885, attempts to weaken its protection led the voters of New York to directly provide even stronger protection. By their approval of an 1894 referendum, New Yorkers added the now-famous "forever wild" clause to the state constitution:

> The lands of the state, now owned or hereafter acquired, constituting the forest preserve as now fixed by law, shall be forever kept as wild forest lands. They shall not be leased, sold or exchanged, or be taken by any corporation, public or private, nor shall the timber thereon be sold, removed or destroyed.[5]

The history of wilderness protection efforts in the Adirondacks deeply influenced wilderness leaders Bob Marshall and, later, Howard Zahniser who cited this constitutional protection as an influence as he sought the strongest possible way to protect federal wilderness areas.[6]

In a 1972 law the New York legislature provided a new plan, resulting in designation of portions of the Adirondack Forest Preserve lands as state wilderness. Today, seventeen areas are designated wilderness, totaling more than 1 million acres. In the same way, four wilderness areas protect more than 100,000 acres in the Catskill Forest Preserve. New York citizen organizations are working to protect additional state wilderness areas.

A New State Wilderness: Minnesota

In a June 2003 law Minnesota designated 18,000 acres of statutory wilderness, activating a state wilderness system created in 1975 legislation that previously applied to no actual lands. The new area comprises lands the state had acquired within the million-acre federal Boundary Waters Canoe Area Wilderness. Protection of these lands is similar to that of federal wilderness. Even as they are pressing for congressional designation of key additions to the federal wilderness, Friends of the Boundary Waters Canoe Area Wilderness and other Minnesota groups are working to gain the same degree of wilderness protection for the larger area of State Trust Lands that remain scattered within the federal wilderness.

The Scale of State-Designated Wilderness

In their 2002 survey, Chad Dawson and Pauline Thorndike found some 2.6 million acres were protected in seventy-four state wilderness areas in seven states, an increase of nearly 60 percent since a similar survey in 1983. Criteria required in the surveys included (1) statutory or administrative recognition of the program, (2) provision for preservation of natural qualities and offering primitive recreation opportunities, and (3) prohibition on development.

Half of the state-designated wilderness areas are smaller than 5,000 acres; eleven are less than 1,000 acres. Twenty are in the 10,000 to 100,000-acre range, and nine are larger. Ninety-six percent of the total acreage is in thirty-four state-designated wilderness areas in Alaska, California, and New York; these average 86,000 acres.

The researchers found that four Midwestern and Eastern states—Michigan, Missouri, Wisconsin, and Maryland—protect forty wilderness areas averaging 2,700 acres in size (the Minnesota area was designated after this survey). Maryland and Michigan's programs were established in statutes modeled on the Wilderness Act.

Wilderness around the World

The United States was the pioneer in recognizing that deliberate government policy is essential to preserve areas of wilderness. A growing number of other nations have been inspired by the Wilderness Act to pursue wilderness preservation policies adapted to their own circumstances. Two experts on international wilderness, Vance Martin and Alan Watson, report that

The concept of wilderness—land and water where natural ecological processes operate as free of human influence as possible, and with primitive recreation opportunities and solitude—has spread from its American roots. Other nations—Australia, New Zealand, Canada, Finland, Sri Lanka, the former Soviet Union, and South Africa, for example—have also legislatively protected wilderness or comparable, strictly protected reserves.[7]

Martin and Watson report that a 1997 database compiled by the World Conservation Monitoring Centre shows that of some 5 million square miles of "protected areas" worldwide, those categorized as "strict nature reserves" and "wilderness" comprised about 302 billion acres. However, such databases are inherently incomplete. Many areas—including some United States wilderness areas—are either not yet in the database or have been sorted into another category. Greenland National Park, for example, is not in either of these subcategories, yet this one area protects some 240 million acres, most of which we would recognize as wilderness in American terms.

It is important to emphasize, as Martin and Watson do, that "the idea of wilderness is slowly gaining recognition and support around the world, with American notions about wilderness an influence, but not a determinant, on the way other nations approach the idea." They stress that "differences in geography, culture, and economics alter the specific ways in which other countries approach the designation and management of wilderness. ..."[8] This variation to meet cultural and other differences is hardly surprising. In the United States, similar special provisions have been made by Congress to deal with unique circumstances of wilderness in Alaska, notably regarding subsistence uses. As these authors stress, "there is increasing acceptance of the term to mean those areas legislated or zoned for protection in their natural condition, and accommodating a wider spectrum of human activity than the U.S. definition might allow."[9]

An elaborate infrastructure of international organizations deals with protection of nature, including the International Union for the Conservation of Nature (IUCN), the World Wide Fund for Nature, and the United Nations Environment Programme, among others. In 1992 the IUCN amended its definitional structure of six categories of protected areas, adding a new subcategory and augmenting its "Strict Nature Reserve" category in order to expressly recognize the worldwide interest in protecting wilderness areas. In this revised international nomenclature, *strict nature preserve* and *wilderness* are distinguished and defined:

CATEGORY Ia Strict Nature Reserve: protected area managed mainly for science

Area of land and/or sea possessing some outstanding or representative ecosystems, geological or physiological features and/or species, available primarily for scientific research and/or environmental monitoring.[10]

CATEGORY Ib Wilderness Area: protected area managed mainly for wilderness protection
Large area of unmodified or slightly modified land, and/or sea, retaining its natural character and influence, without permanent or significant habitation, which is protected and managed so as to preserve its natural condition.[11]

The laws in each nation—or, in some cases, in some of their political subunits—vary a great deal. The degree of security for designated areas varies, too, notably because in parliamentary systems laws can be readily changed when a new party or coalition assumes executive power backed by a legislative majority.

People Saving Wilderness around the World

Vance Martin and Alan Watson highlight perhaps the most far-reaching and lasting influence of the Wilderness Act and the American wilderness movement:

> One of the most important contributions of the American experiment in wilderness preservation has been to demonstrate the powerful influence of public opinion and public involvement. As we look around the world, we find that much of the initiative for wilderness protection everywhere has come from citizen advocacy groups. ... As the millennium turns, we are also realizing that more than organized advocacy is needed—the involvement of nearby residents is crucial to any successful campaign for wilderness designation—whether in the United States or developing nations.[12]

As I was writing this chapter, *The Namibian* of Windhoek, Namibia, reported that the cabinet of Namibia had given its approval to a new national park, comprising a major portion of the Sperrgebiet, a diamond mining reserve in southwestern Namibia. Discovery of diamonds in the area led to barring of public access in German colonial times. The government announcement said the Sperrgebiet is one of the world's top twenty-five biodiversity hot spots and "Namibia, therefore, has an obligation to protect this area as part of its commitments under the UN Convention on Biological Diversity, to which Namibia is a party." According to the news coverage "the government report said the Sperrgebiet fitted the definition of a wilderness 'as an area where the earth and its community of life are untrammeled by man, where man himself is a visitor who does not remain.'"[13]

Howard Zahniser's words, adopted as the policy of the United States four decades ago, echo now around the globe.

Chapter 10

An Enduring Resource of Wilderness

In a delicate balance of idealism and practicality, detailed direction and realistic flexibility, the forty-year-old Wilderness Act in practice has proved to be the vehicle to achieve what Bob Marshall dreamed of: "as close an approximation to permanence as could be realized in a world of shifting desires."[1] As Harvey Broome said, the concept of wilderness has become what it was not before 1964: "an imperative in American life."[2]

We see evidence of the enormous popularity of wilderness and wild adventure all around us. On a recent trip I read about one aspect of the wilderness idea in the in-flight magazine. Writing about his feelings having climbed to the top of Washington's Mount Adams—in the Mount Adams Wilderness Area—Richard Bangs enthuses, in words that could have been John Muir's, "it seemed to be earth's first morning, and we the first to bathe in its beauty, unbounded by geography or history."[3]

Look at this phenomenon from another, very different angle. To sell SUVs, advertising agencies and manufacturers adroitly link their products to Americans' love of wilderness. In the foreground of one ad, an SUV gleams amidst sagebrush and pines on a ridge below towering mountain peaks. (To their credit, the SUV is clearly on a minimal road.) A young family with their knapsacks is striding away into the seemingly endless wild landscape— under a headline expressing wry astonishment that for this they are missing a popular television program. The copy enthuses that, thanks to the capabilities and size of that vehicle, the family will experience nature the way it was intended to be seen—with no commercials.

Big business does not spend tens of millions of dollars on such advertising—for SUVs, rugged outdoor gear marketed to an increasingly urban society, and the like—without intensive consumer research. They know what sells. They know that wilderness taps into something fundamental and positive in the imaginations of the American people.

Wilderness is wildly popular. Its appeal—and the need to preserve more of it—touches most Americans in many ways. At the deepest level is a very broadly shared consensus that we simply owe a generous legacy of wildness to the future. Looked at another way, there is a sense that we have transformed so much of nature that it is increasingly important to save a decent sampling of the original American earth, without succumbing to our own short-term manipulative ambitions. In 1963 the Republican delegations from the eleven western states asked Representative John Saylor to explain an eastern Republican's view of federal land issues. Saylor minced no words:

I see no equity for the nation in insisting that the Western States be allowed to make precisely the same stupid mistakes that were made in the East. This is not, in the long run, equity for the people of the Western States, and it certainly is not equity for the people of the United States.[4]

Saylor went on to explain his long championship for the Wilderness Act:

This is conservation—pure and simple—but it is more than conservation. It is a recognition of the need of mankind—of urbanized mankind—to have places to which those who love solitude can get away for their own good and for the good of all of us. Macadam is good, and airplanes are good, and jangling telephones are good, and apartment house dwelling is good—but they are not good enough. They have to be complemented with roadless areas, with places that are far removed from our cities, and with places where the few who are temporarily within them are far removed from each other.

While the federal agencies do not all keep such statistics, experts have extrapolated that something like 13 million visits were made to wilderness areas in each recent year—a use that is growing. But whether this number is accurate or not, counting "visits" as stepping across a boundary into a designated wilderness area, it hugely undercounts the real *use* of our wilderness areas.

- Wilderness is valued and used in many ways by millions who have never set foot in a wilderness area and never will. I have recently been working with a group of agency and academic experts who have compiled the latest research on these many values; their findings are being published in a forthcoming book, *The Multiple Values of Wilderness.*
- Millions visit and use the unprotected wilderness—and most will be driven from those lands if and when they are developed or overrun with motors and wheels, which will further increase demand for visits to remaining wilderness. This is missed in the official estimates.

This generation is speedily using up, beyond recall, a very important right that belongs to future generations—the right to have wilderness in their civilization, even as we have it in ours; the right to find solitude somewhere; the right to see, and enjoy, and be inspired and renewed, somewhere, by those places where the land of God has not been obscured by the industry of man.

—David Brower,
"Wilderness-Conflict and Conscience,"
Wildlands in Our Civilization

- There are inestimable legions of future users of wilderness—if we save it for them. As Teddy Roosevelt stressed, we have a moral duty "to the number within the womb of time, compared to which those now alive form but an insignificant fraction."

The work of saving America's wilderness—being good stewards to what has been protected and securing additional designations to build the National Wilderness Preservation System—goes on and will go on for generations. As a generalization, in many cases designation of new areas will be more politically challenging for several reasons:

- Some of the most compelling and most threatened areas that should be protected as wilderness are located in places where the current congressional delegations are not particularly sympathetic—and they have great influence in these decisions. As I write this, Alaska and Utah are cases in point.

- Many new proposals involve lower-elevation lands. These are important to protect greater biological diversity and essential wildlife habitat, as well as to provide more all-season wilderness recreational opportunities, but they also often involve more nonconforming uses and human impacts that add to the political complexities that must be resolved.

- The very success of the wilderness advocacy movement has generated greater pushback from other interests. The future use of areas on the federal lands is very often hotly debated.

- An unfortunate degree of partisanship bleeds over from our wider national politics and threatens the long bipartisan tradition in wilderness preservation. Some right-wing think tanks have embraced wilderness opposition as part of their antigovernment agenda. Artfully dressed-up coalitions have been formed to work for "access," a recent example being a coalition of the American Loggers Council, the Wyoming Mining Association, the New Mexico Oil and Gas Association, Citizens Advocating Local Control of Our Forests, and others. A "Wilderness Act Reform Coalition" sputters to life periodically, with ideas for "reforms" that no friend of wilderness preservation would recognize.

- Such interests periodically seek to breathe life into the idea for some kind of alternative system, claiming it would be "just like wilderness," but allowing whatever their particular interests seeks, such as "healthy forest" logging or use of off-road vehicles.

Faced with new challenges, the wilderness advocacy movement must continue to be smart and quick on its feet, building on what we know works: organizing intensive grassroots support for wilderness, constructed on the passionate commitment of an ever growing constituency of unpaid volunteer citizen-activists. After the 1956 Echo Park victory, David Brower spoke of "all the people I saw working so hard for the same thing ... more than we'll ever know, [who] were writing the letters and showing the pictures and riding the river and telling the other people who wrote still more letters and talked to still more people all of whom, in the nameless but undeniable aggregate, chalked up the victory."[5]

As other forces strive to undercut public support for wilderness preservation, wilderness advocates must remember this: as important as the Wilderness Act is, the law itself is only a piece of paper covered with words; the original copy, printed on vellum and signed by President Johnson, is enshrined in the National Archives. In the final analysis, it is not the law that protects the wilderness areas we have designated, it is the social consensus the law represents that keeps it strong. We who love wilderness must not take that social consensus for granted.

I firmly believe wilderness has its own inherent right to exist, with its community of life "pasturing freely where we never wander."[6] When I speak of preserving these remarkable wild lands for future generations, I am thinking not just of future generations of *us*, but of the grandchildren's grandchildren of the grizzlies and wild salmon, too. Nonetheless, we preserve wilderness best and serve that noble purpose best when we remember and stress that it is people who preserve wilderness—for people. Ordinary people no different from you and me worked, year after year, to protect wilderness and to secure enactment of the federal laws that further that goal. People just like us are ever vigilant to see that the wilderness we have designated is well administered—and that development threats to potential wilderness areas are held at bay until Congress can act. And people like us are the force that will repel efforts to weaken protection or to remove designated wilderness from its statutory protection—or to "reform" the Wilderness Act.

This is not work that can be successfully completed by confrontational zealots, though wilderness advocates must and do exhibit extraordinary zeal. Nor is it suitable work for cynics ever eager to trash the democratic processes through which our diverse and pluralistic society reaches and honors its collective decisions. Progress for wilderness has been made by those with moderate and constructive voices, those who appreciate how our government works and learn how to effectively work within it. For his recently completed dissertation at Princeton, historian Jay Turner studied the politics of the additional wilderness designations in the first three decades following enactment of the Wilderness Act. He concluded that to the extent some now focus on more confrontational or uncompromising approaches to wilderness advocacy, they have risked abandoning the "more moderate and pragmatic tradition of wilderness advocacy that dates back to the Wilderness Act of 1964."

From time to time, there have been those within the wilderness movement who envision shortcuts around the realities of legislative politics. They become quite critical of the incremental approach to designation of wilderness and the practical accommodations that wilderness leaders in many states have made to work out wilderness designation proposals that can be enacted. Much of this strategic disagreement is carried on at a constructive and useful level in which alternative visions and strategies are honestly and respectfully discussed. However, as in any passionate social movement, some disagreements descend into harsher debate. A common theme of some critics is that wilderness advocates should not countenance legislative incrementalism and accommodations, but hold out for a grander vision.

One wilderness leader who has contended with criticism of this sort is my friend Rick Johnson, director of the Idaho Conservation League. Rick keeps handy a pertinent reminder from Theodore Roosevelt, who certainly knew what passionate political disputes could be like.

It is not the critic who counts; not the man who points out how the strong man stumbles, or where the doer of deeds could have done them better. The credit belongs to the man who is actually in the arena, whose face is marred by dust and sweat and blood, who strives valiantly; who errs and comes up short again and again; because there is not effort without error and shortcomings; but who does actually strive to do the deed; who knows the great enthusiasm, the great devotion, who spends himself in a worthy cause, who at the best knows in the end the triumph of high achievement and who at the worst, if he fails, at least he fails while daring greatly, so that his place shall never be with those cold and timid souls who know neither victory nor defeat.[7]

Recently a group of third-generation ranchers from the spectacularly wild Rocky Mountain Front country south of Glacier National Park traveled to Washington, D.C., to urge protection of national forest wild lands near their ranches and abutting existing designated wilderness, lands gravely threatened by oil and gas drilling. When they met with Montana's Republican Senator Conrad Burns, a reporter tagged along and reported that the senator seemed unmoved by their advocacy. "I just can't bring myself to believe they will drill the Front," he told the ranchers, "and if they do they will comply with the law." Asked for his reaction, rancher Karl Rappold told the reporter he felt neither that he had been bamboozled nor that he had wasted his time. "Just because he's on one side of the fence and we are on the other, doesn't mean we won't find a gateway."[8]

The fate of unmodified Nature rests in the activity of its friends. If they continue to be too busy or too indifferent to unite in its defense, then the universe of the wilderness is doomed to early extinction. If, on the other hand, they believe that its preservation is worth the sacrifice of some precious time and energy, and if they will take the trouble to become vociferous, there is no reason why material areas of America should not be kept primitive forever.

—Robert Marshall,
"The Universe of the Wilderness Is Vanishing,"
Nature Magazine, April 1937

A rancher, not a hard-bitten lobbyist or politician, put his finger on the key to successful wilderness preservation. Choosing up sides and having pitched battles is sometimes the only choice, but inevitably it always comes down to working things out in some degree of accommodation—as it should in the government of a diverse and pluralistic society.

This is the lesson of how the National Wilderness Preservation System has been built over four decades. This is where you and I can get engaged—and are needed. Active grassroots wilderness organizations and coalitions are at work on behalf of new wilderness designation proposals in every state with public lands. They are eager for your help—doing field studies (lots of fun); assembling supportive information (if you like research); enlisting and mobilizing local public support, particularly among civic leaders, business owners, ranchers, hunters and anglers, and other recreationists (great way to be neighborly); working with local media (your name in lights!); helping evolve strategy (all ideas welcome); and working with agency and congressional staffs and members of Congress (practicing democracy, the surest path to real patriotism I know). There are roles to suit anyone; all are important and more hands, brains, and voices are needed. To find wilderness groups in your own state, check out www.leaveitwild.org.

While your activism will have greatest impact in your own state, there is a need for nationwide voices to be raised whenever potential wilderness is under threat— places such as Montana's Rocky Mountain Front, the superb wild lands of New England, threatened forests in the South and across the West, the vast and imperiled redrock country of southern Utah, the coastal plain of the Arctic National Wildlife Refuge, and the rainforests of southeast Alaska.

To this work, everyday citizens like rancher Karl Rappold—and like you, reader—bring what is most important: your commitment to the heroic ideal of our participatory democracy. I often think the greatest enemy of wilderness preservation is the corrosive cynicism so pervasive in our society. Nothing can be more important than persuading our fellow citizens, and particularly youth, of the self-fulfilling nature of political cynicism. My friend Professor Kai Lee at Williams College reminded me recently of why wilderness is preserved by heroes just like you. "A democracy works," he told a senior seminar where he had asked me to speak, "only if politics is the avocation of a large fraction of the citizenry." And he cited the German sociologist Max Weber:

> Politics is a strong and slow boring of hard boards. It takes both passion
> and perspective. Certainly all historical experience confirms the truth—that
> man would not have attained the possible unless time and again he had
> reached out for the impossible. But to do that a man must be a leader, and
> not only a leader but a hero as well, in a very sober sense of the word. And
> even those who are neither leaders nor heroes must arm themselves with
> that steadfastness of heart which can brave even the crumbling of all hopes.[9]

Looking back at the history of wilderness preservation, I know such citizen heroism works. I know that committed grassroots citizen-activists can and do save wilderness, ordinary people devoted to protecting their favorite wild lands and willing to undertake, the "strong and slow boring of hard boards." I know that trying to "find a gateway," even with those who seem most hostile or indifferent, works. I also know that nothing else saves our wilderness.

It seems to me that Aldo Leopold, who said that wilderness is a "fundamental instrument for building citizens,"[10] was saying something profound, and in more than one sense.

For one thing, Leopold was certainly reminding us of the essential influence of wilderness in the innate American character, planted lastingly there by our national history. Frederick Jackson Turner said as much in a commencement address in 1914: "The appeal of the undiscovered is strong in America. For three centuries the fundamental process in its history was the westward movement, the discovery and occupation of the vast free spaces of the continent."[11] He went on to say that those pioneers were unaware

> of the fact that their most fundamental traits, their institutions, even their ideals were shaped by this interaction between the wilderness and themselves. American democracy was born of no theorist's dream; it was not carried in the *Sarah Constant* to Virginia, nor in the *Mayflower* to Plymouth. It came out of the American forest, and it gained new strength each time it touched a new frontier.

With the enactment of the Wilderness Act, Leopold's notion of wilderness as "a fundamental instrument for building citizens" took on a second dimension. In the very act of working to save wilderness, thousands upon thousands of ordinary, everyday citizens have learned, practiced, and honored our democratic traditions of civic engagement. At the frontiers of today's wilderness movement, our democracy is gaining strength still. Little wonder former senators Dale Bumpers and Dan Evans extolled this process as the most lowercase-d democratic of land-use decisions.

Passion. Boldness. Steadfastness. A commitment to understanding and participation in our democracy, fueled not only by commitment to wilderness but by love of our participatory democracy itself. These are among the qualities required of those who save wilderness. Far from being just one segment of outdoor users competing with other kinds of outdoor recreationists, wilderness protectors are fired by something much deeper. At bottom, I think, is a passionately felt sense of moral obligation to the people of the future. No one has voiced this more profoundly than Terry Tempest Williams:

> The eyes of the future are looking back at us and they are praying for us to see beyond our own time. Our descendants are kneeling with clasped hands

hoping that we might act with restraint, that we might leave room for the
life that is destined to come. To protect what is wild is to protect what is
gentle. Perhaps the wilderness we fear is the pause between our own
heartbeats, the silence that reminds us we live only by grace. Wilderness
lives by this same grace. It is within our legislative power to create
merciful laws.[12]

Readers, you will know that I deeply revere Howard Zahniser. Though I never
knew him, as a graduate student decades ago I adopted him as my wilderness and
political mentor. So it is Zahnie who will have the last word here. At a Sierra Club
wilderness conference in 1961, he looked into the future; in prophetic words he
reminded us that he and I—and you—are engaged:

> Working to preserve in perpetuity is a great inspiration. We are not
> fighting a rear-guard action, we are facing a frontier. We are not slowing
> down a force that inevitably will destroy all the wilderness there is. We are
> generating another force, never to be wholly spent, that, renewed
> generation after generation, will be always effective in preserving
> wilderness. We are not fighting progress. We are making it. We are not
> dealing with a vanishing wilderness. We are working for a wilderness
> forever.[13]

Afterword

Henry Thoreau wrote in the 1850s that every town and village should have its own wilderness of 500 or 1,000 acres in its midst. All these years later, go to Morristown, New Jersey, and take Thoreau along with you. Get wildly lost with him in the wilderness of the Great Swamp National Wildlife Refuge. This beautiful wild space is protected by the Wilderness Act and yet is surrounded by suburbia just twenty-six miles from Times Square. Imagine the smile on Henry's face!

The Wilderness Act was founded on just this sort of wild imagination, rooted in the 1840s and 1850s. This imagination has been nurtured by fits and starts right up to the present. As we carry on this work, this wild imagination inspires us as advocates of the experience of nature, of the more-than-human world.

Thus, the Wilderness Act was not merely the product of an eight-year struggle for a piece of legislation. It took more than 150 years of sustained wild imagination to establish a National Wilderness Preservation System. If you are part of this movement as an activist or wilderness steward—perhaps even the newest recruit—you are part of a collective imagination transcending our own mortal life spans. You take part in a multigenerational hope that can sustain us through the most disheartening times. You are surrounded by a host of inspired conservationists that preceded you as you go about your own work of protecting nature, wildness, the more-than-human world.

My father, Howard Zahniser, said we must always keep in mind that the essential character of wilderness is its wildness. The great naturalist John Hay mirrored that sentiment in saying that wilderness is not just designated areas but "the Earth's immortal genius." Further, the scientist Lynn Margulis says there are more cells of other beings in and on your body than there are cells of your own. We not only inhabit the wild, in other words, the wild inhabits us.

To preserve wilderness in perpetuity is to sustain the wildness that is its essence, the will of the land itself. My father felt this most strongly. By preserving the wilderness, he wrote, we project into the eternity of the future some of that precious, unspoiled ecological inheritance that has come down to us out of the eternity of the past. Doug Scott extends the thought—that perpetuity is not just about protecting wilderness for our grandchildren's grandchildren. As importantly, preservation of wilderness is also about the grandchildren's grandchildren of the grizzly, and of the spotted owl, and of the wild salmon.

What will your own wild imagination add to this great tradition? Remember this: you are Thoreau. You are Hay. You are Bob Marshall, and Rachel Carson, and

Aldo Leopold, and Mardy Murie, and John Muir. Taking your advocacy for wilderness into the halls of Congress, you are Howard Zahniser. All of them had great hopes for you. Build on that hope by adding to their legacy.

—Ed Zahniser

Appendix

The Wilderness Act and Related Laws

The full text of the Wilderness Act and two related statutes appears below. More than 110 additional laws have been enacted since 1964 adding lands to the National Wilderness Preservation System, and many of these have important provisions and legislative history relating to administration of those wilderness areas.[1]

The best source for gaining a comprehensive understanding of this body of wilderness law is *The Wilderness Act Handbook* published by The Wilderness Society. The thoroughly updated 40th anniversary edition may be downloaded from the Web at www.wilderness.org/Library/Documents/WildernessActHandbook2004.cfm.

Copies can also be ordered by mail for $5.00 (includes shipping and handling) from Publications Office, The Wilderness Society, 1615 M Street, NW, Washington, DC 20036.

Explanations of the background and legislative history concerning specific topics, such as mechanization and wilderness, are found in a series of Briefing Papers on the Campaign for America's Wilderness Web site at http://leaveitwild.org/reports/briefing.html. Additional papers, as well as updates on those already there, are added frequently. I am also happy to respond to questions: dscott@leaveitwild.org.

—Doug Scott

Wilderness Act of 1964

Public Law 88-577 (16 U.S. C. 1131-1136)
88th Congress, Second Session
September 3, 1964
AN ACT

---

[1] The text of every wilderness designation law can be found in the public law library at www.wilderness.net. They are most readily found by searching under the name of each wilderness area or state (all relevant laws will be found this way). Wilderness.net, an invaluable Web resource, is a partnership project of the Wilderness Institute at the University of Montana's College of Forestry and Conservation, the Arthur Carhart National Wilderness Training Center, and the Aldo Leopold Wilderness Research Institute (the latter two are joint programs of the four federal wilderness agencies).

To establish a National Wilderness Preservation System for the permanent good of the whole people, and for other purposes.

Be it enacted by the Senate and House of Representatives of the United States of America in Congress assembled,

Short Title

SEC. 1. This Act may be cited as the "Wilderness Act".

Wilderness System Established Statement of Policy

SEC. 2. (a) In order to assure that an increasing population, accompanied by expanding settlement and growing mechanization, does not occupy and modify all areas within the United States and its possessions, leaving no lands designated for preservation and protection in their natural condition, it is hereby declared to be the policy of the Congress to secure for the American people of present and future generations the benefits of an enduring resource of wilderness. For this purpose there is hereby established a National Wilderness Preservation System to be composed of federally owned areas designated by Congress as "wilderness areas", and these shall be administered for the use and enjoyment of the American people in such manner as will leave them unimpaired for future use and enjoyment as wilderness, and so as to provide for the protection of these areas, the preservation of their wilderness character, and for the gathering and dissemination of information regarding their use and enjoyment as wilderness; and no Federal lands shall be designated as "wilderness areas" except as provided for in this Act or by a subsequent Act.

(b) The inclusion of an area in the National Wilderness Preservation System notwithstanding, the area shall continue to be managed by the Department and agency having jurisdiction thereover immediately before its inclusion in the National Wilderness Preservation System unless otherwise provided by Act of Congress. No appropriation shall be available for the payment of expenses or salaries for the administration of the National Wilderness Preservation System as a separate unit nor shall any appropriations be available for additional personnel stated as being required solely for the purpose of managing or administering areas solely because they are included within the National Wilderness Preservation System.

Definition of Wilderness

(c) A wilderness, in contrast with those areas where man and his own works dominate the landscape, is hereby recognized as an area where the earth and its community of life are untrammeled by man, where man himself is a visitor who does not remain. An area of wilderness is further defined to mean in this Act an area of undeveloped Federal land retaining its primeval character and influence, without permanent improvements or human habitation, which is protected and managed so as to preserve its natural conditions and which (1) generally appears to have been affected primarily by the forces of nature, with the imprint of man's work substantially unnoticeable; (2) has outstanding opportunities for solitude or a primitive and unconfined type of recreation; (3) has at least five thousand acres of land or is of sufficient size as to make practicable its preservation and use in an unimpaired condition; and (4) may also contain ecological, geological, or other features of scientific, educational, scenic, or historical value.

National Wilderness Preservation System—Extent of System

SEC. 3. (a) All areas within the national forests classified at least 30 days before the effective date of this Act by the Secretary of Agriculture or the Chief of the Forest Service as "wilderness", "wild", or "canoe" are hereby designated as wilderness areas. The Secretary of Agriculture shall-

(1) Within one year after the effective date of this Act, file a map and legal description of each wilderness area with the Interior and Insular Affairs Committees of the United States Senate and the

House of Representatives, and such descriptions shall have the same force and effect as if included in this Act: *Provided, however,* That correction of clerical and typographical errors in such legal descriptions and maps may be made.

(2) Maintain, available to the public, records pertaining to said wilderness areas, including maps and legal descriptions, copies of regulations governing them, copies of public notices of, and reports submitted to Congress regarding pending additions, eliminations, or modifications. Maps, legal descriptions, and regulations pertaining to wilderness areas within their respective jurisdictions also shall be available to the public in the offices of regional foresters, national forest supervisors, and forest rangers.

(b) The Secretary of Agriculture shall, within ten years after the enactment of this Act, review, as to its suitability or nonsuitability for preservation as wilderness, each area in the national forests classified on the effective date of this Act by the Secretary of Agriculture or the Chief of the Forest Service as "primitive" and report his findings to the President. The President shall advise the United States Senate and House of Representatives of his recommendations with respect to the designation as "wilderness" or other reclassification of each area on which review has been completed, together with maps and a definition of boundaries. Such advice shall be given with respect to not less than one-third of all the areas now classified as "primitive" within three years after the enactment of this Act, not less than two-thirds within seven years after the enactment of this Act, and the remaining areas within ten years after the enactment of this Act. Each recommendation of the President for designation as "wilderness" shall become effective only if so provided by an Act of Congress. Areas classified as "primitive" on the effective date of this Act shall continue to be administered under the rules and regulations affecting such areas on the effective date of this Act until Congress has determined otherwise. Any such area may be increased in size by the President at the time he submits his recommendations to the Congress by not more than five thousand acres with no more than one thousand two hundred and eighty acres of such increase in any one compact unit; if it is proposed to increase the size of any such area by more than five thousand acres or by more than one thousand two hundred and eighty acres in any one compact unit the increase in size shall not become effective until acted upon by Congress. Nothing herein contained shall limit the President in proposing, as part of his recommendations to Congress, the alteration of existing boundaries of primitive areas or recommending the addition of any contiguous area of national forest lands predominantly of wilderness value. Notwithstanding any other provisions of this Act, the Secretary of Agriculture may complete his review and delete such area as may be necessary, but not to exceed seven thousand acres, from the southern tip of the Gore Range-Eagles Nest Primitive Area, Colorado, if the Secretary determines that such action is in the public interest.

(c) Within ten years after the effective date of this Act the Secretary of the Interior shall review every roadless area of five thousand contiguous acres or more in the national parks, monuments and other units of the national park system and every such area of, and every roadless island within, the national wildlife refuges and game ranges, under his jurisdiction on the effective date of this Act and shall report to the President his recommendation as to the suitability or nonsuitability of each such area or island for preservation as wilderness. The President shall advise the President of the Senate and the Speaker of the House of Representatives of his recommendation with respect to the designation as wilderness of each such area or island on which review has been completed, together with a map thereof and a definition of its boundaries. Such advice shall be given with respect to not less than one-third of the areas and islands to be reviewed under this subsection within three years after enactment of this Act, not less than two-thirds within seven years of enactment of this Act, and the remainder within ten years of enactment of this Act. A recommendation of the President for designation as wilderness shall become effective only if so provided by an Act of Congress. Nothing contained herein shall, by implication or otherwise, be construed to lessen the present statutory authority of the Secretary of the Interior with respect to the maintenance of roadless areas within units of the national park system.

(d) (1) The Secretary of Agriculture and the Secretary of the Interior shall, prior to submitting any recommendations to the President with respect to the suitability of any area for preservation as wilderness-

(A) give such public notice of the proposed action as they deem appropriate, including publication in the Federal Register and in a newspaper having general circulation in the area or areas in the vicinity of the affected land;

(B) hold a public hearing or hearings at a location or locations convenient to the area affected. The hearings shall be announced through such means as the respective Secretaries involved deem appropriate, including notices in the Federal Register and in newspapers of general circulation in the area: *Provided*, That if the lands involved are located in more than one State, at least one hearing shall be held in each State in which a portion of the land lies;

(C) at least thirty days before the date of a hearing advise the Governor of each State and the governing board of each county, or in Alaska the borough, in which the lands are located, and Federal departments and agencies concerned, and invite such officials and Federal agencies to submit their views on the proposed action at the hearing or by no later than thirty days following the date of the hearing.

(2) Any views submitted to the appropriate Secretary under the provisions of (1) of this subsection with respect to any area shall be included with any recommendations to the President and to Congress with respect to such area.

(e) Any modification or adjustment of boundaries of any wilderness area shall be recommended by the appropriate Secretary after public notice of such proposal and public hearing or hearings as provided in subsection (d) of this section. The proposed modification or adjustment shall then be recommended with map and description thereof to the President. The President shall advise the United States Senate and the House of Representatives of his recommendations with respect to such modification or adjustment and such recommendations shall become effective only in the same manner as provided for in subsections (b) and (c) of this section.

Use of Wilderness Areas

SEC. 4. (a) The purposes of this Act are hereby declared to be within and supplemental to the purposes for which national forests and units of the national park and national wildlife refuge systems are established and administered and-

(1) Nothing in this Act shall be deemed to be in interference with the purpose for which national forests are established as set forth in the Act of June 4, 1897 (30 Stat. 11), and the Multiple Use Sustained-Yield Act of June 12, 1960 (74 Stat. 215).

(2) Nothing in this Act shall modify the restrictions and provisions of the Shipstead-Nolan Act (Public Law 539, Seventy first Congress, July 10, 1930; 46 Stat. 1020), the Thye-Blatnik Act (Public Law 733, Eightieth Congress, June 22, 1948; 62 Stat. 568), and the Humphrey-Thye-Blatnik-Andresen Act (Public Law 607, Eighty-fourth Congress, June 22, 1956; 70 Stat. 326), as applying to the Superior National Forest or the regulations of the Secretary of Agriculture.

(3) Nothing in this Act shall modify the statutory authority under which units of the national park system are created. Further, the designation of any area of any park, monument, or other unit of the national park system as a wilderness area pursuant to this Act shall in no manner lower the standards evolved for the use and preservation of such park, monument, or other unit of the national park system in accordance with the Act of August 25, 1916, the statutory authority under which the area was created, or any other Act of Congress which might pertain to or affect such area, including, but not limited to, the Act of June 8, 1906 (34 Stat. 225; 16 U.S.C. 432 et seq.); section 3(2) of the Federal Power Act (16 U.S.C. 796(2)): and the Act of August 21, 1935 (49 Stat. 666; 16 U.S.C. 461 et seq.).

(b) Except as otherwise provided in this Act, each agency administering any area designated as wilderness shall be responsible for preserving the wilderness character of the area and shall so administer such area for such other purposes for which it may have been established as also to preserve its wilderness character. Except as otherwise provided in this Act, wilderness areas shall be devoted to the public purposes of recreational, scenic, scientific, educational, conservation, and historical use.

Prohibition of Certain Uses

(c) Except as specifically provided for in this Act, and subject to existing private rights, there shall be no commercial enterprise and no permanent road within any wilderness area designated by this Act and, except as necessary to meet minimum requirements for the administration of the area for the purpose of this Act (including measures required in emergencies involving the health and safety of persons within the area), there shall be no temporary road, no use of motor vehicles, motorized equipment or motorboats, no landing of aircraft, no other form of mechanical transport, and no structure or installation within any such area.

Special Provisions

(d) The following special provisions are hereby made:

(1) Within wilderness areas designated by this Act the use of aircraft or motorboats, where these uses have already become established, may be permitted to continue subject to such restrictions as the Secretary of Agriculture deems desirable. In addition, such measures may be taken as may be necessary in the control of fire, insects, and diseases, subject to such conditions as the Secretary deems desirable.

(2) Nothing in this Act shall prevent within national forest wilderness areas any activity, including prospecting, for the purpose of gathering information about mineral or other resources, if such activity is carried on in a manner compatible with the preservation of the wilderness environment. Furthermore, in accordance with such program as the Secretary of the Interior shall develop and conduct in consultation with the Secretary of Agriculture, such areas shall be surveyed on a planned, recurring basis consistent with the concept of wilderness preservation by the Geological Survey and the Bureau of Mines to determine the mineral values, if any, that may be present; and the results of such surveys shall be made available to the public and submitted to the President and Congress.

(3) Notwithstanding any other provisions of this Act, until midnight December 31, 1983, the United States mining laws and all laws pertaining to mineral leasing shall, to the same extent as applicable prior to the effective date of this Act, extend to those national forest lands designated by this Act as "wilderness areas"; subject, however, to such reasonable regulations governing ingress and egress as may be prescribed by the Secretary of Agriculture consistent with the use of the land for mineral location and development and exploration, drilling, and production, and use of land for transmission lines, waterlines, telephone lines, or facilities necessary in exploring, drilling, producing, mining, and processing operations, including where essential, the use of mechanized ground or air equipment and restoration as near as practicable of the surface of the land disturbed in performing prospecting, location, and, in oil and gas leasing, discovery work, exploration, drilling, and production, as soon as they have served their purpose. Mining locations lying within the boundaries of said wilderness areas shall be held and used solely for mining or processing operations and uses reasonably incident thereto; and hereafter, subject to valid existing rights, all patents issued under the mining laws of the United States affecting national forest lands designated by this Act as wilderness areas shall convey title to the mineral deposits within the claim, together with the right to cut and use so much of the mature timber therefrom as may be needed in the extraction, removal, and beneficiation of the mineral deposits, if needed timber is not otherwise reasonably available, and if the timber is cut under sound principles of forest management as defined by the national forest rules and regulations, but each such patent shall reserve to the United States all title in or to the surface of the lands and products thereof, and no use of the surface of the claim or the resources therefrom not reasonably required for carrying on mining or prospecting shall be allowed except as otherwise expressly provided in this Act:

Provided, That, unless hereafter specifically authorized, no patent within wilderness areas designated by this Act shall issue after December 31, 1983, except for the valid claims existing on or before December 31, 1983. Mining claims located after the effective date of this Act within the boundaries of wilderness areas designated by this Act shall create no rights in excess of those rights which may be patented under the provisions of this subsection. Mineral leases, permits, and licenses covering lands within national forest wilderness areas designated by this Act shall contain such reasonable stipulations as may be prescribed by the Secretary of Agriculture for the protection of the wilderness character of the land consistent with the use of the land for the purposes for which they are leased, permitted, or licensed. Subject to valid rights then existing, effective January 1, 1984, the minerals in lands designated by this Act as wilderness areas are withdrawn from all forms of appropriation under the mining laws and from disposition under all laws pertaining to mineral leasing and all amendments thereto.

(4) Within wilderness areas in the national forests designated by this Act, (1) the President may, within a specific area and in accordance with such regulations as he may deem desirable, authorize prospecting for water resources, the establishment and maintenance of reservoirs, water-conservation works, power projects, transmission lines, and other facilities needed in the public interest, including the road construction and maintenance essential to development and use thereof, upon his determination that such use or uses in the specific area will better serve the interests of the United States and the people thereof than will its denial; and (2) the grazing of livestock, where established prior to the effective date of this Act, shall be permitted to continue subject to such reasonable regulations as are deemed necessary by the secretary of Agriculture.

(5) Other provisions of this Act to the contrary notwithstanding, the management of the Boundary Waters Canoe Area, formerly designated as the Superior, Little Indian Sioux, and Caribou Roadless Areas, in the Superior National Forest. Minnesota, shall be in accordance with regulations established by the Secretary of Agriculture in accordance with the general purpose of maintaining, without unnecessary restrictions on other uses, including that of timber, the primitive character of the area, particularly in the vicinity of lakes, streams, and portages: *Provided*, That nothing in this Act shall preclude the continuance within the area of any already established use of motorboats.[2]

(6) Commercial services may be performed within the wilderness areas designated by this Act to the extent necessary for activities which are proper for realizing the recreational or other wilderness purposes of the areas.

(7) Nothing in this Act shall constitute an express or implied claim or denial on the part of the Federal Government as to exemption from State water laws.

(8) Nothing in this Act shall be construed as affecting the jurisdiction or responsibilities of the several States with respect to wildlife and fish in the national forests.

State and Private Lands within Wilderness Areas

SEC. 5. (a) In any case where State-owned or privately owned land is completely surrounded by national forest lands within areas designated by this act as wilderness, such State or private owner shall be given such rights as may be necessary to assure adequate access to such State-owned or privately owned land by such State or private owner and their successors in interest, or the State-owned land or privately owned land shall be exchanged for federally owned land in the same State of approximately equal value under authorities available to the Secretary of Agriculture: *Provided, however*, That the United States shall not transfer to a State or private owner any mineral interests unless the State or private owner relinquishes or causes to be relinquished to the United States the mineral interest in the surrounded land.

---

[2] This paragraph was repealed by Congress when it enacted new legislation renaming the area as the "Boundary Waters Canoe Area Wilderness," expanding its boundaries and reforming some aspects of its prior management to better protect wilderness values.

(b) In any case where valid mining claims or other valid occupancies are wholly within a designated national forest wilderness area, the Secretary of Agriculture shall, by reasonable regulations consistent with the preservation of the area as wilderness, permit ingress and egress to such surrounded areas by means which have been or are being customarily enjoyed with respect to other such areas similarly situated.

(c) Subject to the appropriation of funds by Congress, the Secretary of Agriculture is authorized to acquire privately owned land within the perimeter of any area designated by this Act as wilderness if (1) the owner concurs in such acquisition or (2) the acquisition is specifically authorized by Congress.

Gifts, Bequests, and Contributions

SEC. 6. (a) The Secretary of Agriculture may accept gifts or bequests of land within wilderness areas designated by this Act for preservation as wilderness. The Secretary of Agriculture may also accept gifts or bequests of land adjacent to wilderness areas designated by this Act for preservation as wilderness if he has given sixty days advance notice thereof to the President of the Senate and the Speaker of the House of Representatives. Land accepted by the Secretary of Agriculture under this section shall become part of the wilderness area involved. Regulations with regard to any such land may be in accordance with such agreements, consistent with the policy of this Act, as are made at the time of such gift, or such conditions, consistent with such policy, as may be included in, and accepted with, such bequest.

(b) The Secretary of Agriculture or the Secretary of the Interior is authorized to accept private contributions and gifts to be used to further the purposes of this Act.

Annual Reports

SEC. 7. At the opening of each session of Congress, the Secretaries of Agriculture and Interior shall jointly report to the President for transmission to Congress on the status of the wilderness system, including a list and descriptions of the areas in the system, regulations in effect, and other pertinent information, together with any recommendations they may care to make.[3]

Congress Adds Lands Administered by the Bureau of Land Management to the Wilderness Program

Even before the Wilderness Act became law, it was apparent that leaving out the vast public domain lands administered by the Bureau of Land Management was a serious gap in the basic architecture of the National Wilderness Preservation System. This anomaly was corrected by the following provisions of a 1976 law.

---

[3] This annual reporting requirement was terminated by Congress in 1995 as part of a law terminating hundreds of such requirements in various federal laws.

Federal Land Policy and Management Act of 1976
Public Law 94-579
(94the Congress, 2nd session)
October 21, 1976
Section 603 is codified as 43 U.S.C. 1782

Bureau of Land Management Wilderness Study

Sec. 603. (a) Within fifteen years after the date of approval of this Act [October 21, 1976], the Secretary shall review those roadless areas of five thousand acres or more and roadless islands of the public lands, identified during the inventory required by section 201(a) of this Act as having wilderness characteristics described in the Wilderness Act of September 3, 1964 (78 Stat. 890; 16 U.S.C. 1131 et seq.) and shall from time to time report to the President his recommendation as to the suitability or nonsuitability of each such area or island for preservation as wilderness: *Provided,* That prior to any recommendations for the designation of an area as wilderness the Secretary shall cause mineral surveys to be conducted by the United States Geological Survey and the United States Bureau of Mines to determine the mineral values, if any, that may be present in such areas: *Provided further,* That the Secretary shall report to the President by July 1, 1980, his recommendations on those areas which the Secretary has prior to November 1, 1975, formally identified as natural or primitive areas. The review required by this subsection shall be conducted in accordance with the procedure specified in section 3(d) of the Wilderness Act.

(b) The President shall advise the President of the Senate and the Speaker of the House of Representatives of his recommendations with respect to designation as wilderness of each such area, together with a map thereof and a definition of its boundaries. Such advice by the President shall be given within two years of the receipt of each report from the Secretary. A recommendation of the President for designation as wilderness shall become effective only if so provided by an Act of Congress.

(c) During the period of review of such areas and until Congress has determined otherwise, the Secretary shall continue to manage such lands according to his authority under this Act and other applicable law in a manner so as not to impair the suitability of such areas for preservation as wilderness, subject, however, to the continuation of existing mining and grazing uses and mineral leasing in the manner and degree in which the same was being conducted on the date of approval of this Act [October 21, 1976]: *Provided,* That, in managing the public lands the Secretary shall by regulation or otherwise take any action required to prevent unnecessary or undue degradation of the lands and their resources or to afford environmental protection. Unless previously withdrawn from appropriation under the mining laws, such lands shall continue to be subject to such appropriation during the period of review unless withdrawn by the Secretary under the procedures of section 204 this Act for reasons other than preservation of their wilderness character. Once an area has been designated for preservation as wilderness, the provisions of the Wilderness Act which apply to national forest wilderness areas shall apply with respect to the administration and use of such designated area, including mineral surveys required by section 4(d)(2) of the Wilderness Act, and mineral development, access, exchange of lands, and ingress and egress for mining claimants and occupants.

Endnotes

Foreword

1 Aldo Leopold, "Wilderness as a Form of Land Use," *The Journal of Land and Public Utility Economics* 1 (October, 1925): 400.

2 Theodore Roosevelt, "New Nationalism" speech, Oswatomie, Kansas, August 31, 1910. For full text of speech, visit http://teachingamericanhistory.org/library/index.asp?document=501.

Introduction

1 Aldo Leopold, "Wilderness as a Form of Land Use," *The Journal of Land and Public Utility Economics* 1 (October, 1925): 400.

2 Howard Zahniser, "The Wilderness Bill and Foresters," March 14, 1957, talk reprinted as part of his testimony, Senate Committee on Interior and Insular Affairs, *National Wilderness Preservation System: Hearings before the Committee on Interior and Insular Affairs*, 85th Cong., 1st sess., June 19–20, 1957, 211.

3 Stewart Udall, testimony, Senate Committee on Interior and Insular Affairs, *National Wilderness Preservation Act: Hearings before the Committee on Interior and Insular Affairs*, 88th Cong., 1st sess., February 28–March 1, 1963, 16.

4 President Gerald R. Ford, Message from the President of the United States, Transmitting the Tenth Annual Report on the Status of the National Wilderness Preservation System, December 4, 1974.

5 Paul C. Light, "Government's Greatest Achievements of the Past Half Century," *Reform Watch* Brief No. 2 (November 2000). http://www.brook.edu/comm/reformwatch/rw02.htm.

6 Senators Dale Bumpers and Dan Evans, op-ed, as distributed by Writers on the Range and first published as "Gone Wild," *Missoula Independent*, May 13, 2004, http://www.missoulanews.com/News/News.asp?no=4046.

7 Susan Snyder, "Congress should use Hall way," *Las Vegas Sun*, March 29, 2004.

8 Neal Rahm, speech to Forest Service staff, March 13, 1971, in records of Robert Morgan, as cited by Dennis M. Roth, *The Wilderness Movement and the National Forests* (College Station, Tex.: Intaglio Press, 1988), 35.

Chapter 1: Wilderness in Our Lives

1 The areas mentioned here are a 12,925,000-acre unit stretching across the Gates of the Arctic National Park and Noatak National Preserve, Alaska; the wilderness areas within the Great Swamp National Wildlife Refuge, New Jersey; the Pelican Island National Wildlife Refuge, Florida; and the Forest Service administered Sandia Wilderness in New Mexico, Mount Naomi Wilderness in Utah, and Pusch Ridge Wilderness in Arizona.

2 Senator Pete Domenici, *Congressional Record*, May 5, 1980, 9895.

3 John C. Hendee and Chad P. Dawson, "Wilderness: Progress after Forty Years Under the U.S. Wilderness Act," *International Journal of Wilderness* (April 2004): 4.

4 Aldo Leopold, *A Sand County Almanac* (New York: Oxford University Press, 1949), 188–189.

5 Provisions of the Clean Air Act provide for the protection of the pristine air quality in those wilderness areas of more than 5,000 acres established prior to August 1977. This "prevention of significant deterioration of air quality" provision is found in the U.S. Code at 42 USC Part C.

6 Representative John P. Saylor, *Congressional Record*, July 12, 1956, page 17 of booklet reprint.

7 House Committee on Interior and Insular Affairs, *Designation of Wilderness Areas: Hearings before the Subcommittee on Public Lands and National Parks and Recreation*, 91st Cong., 2nd sess., June 26, 1970, 501.

8 Aldo Leopold, "Conserving the Covered Wagon," *Sunset* (March 1925): 56.

9 Senate Committee on Interior and Insular Affairs, *Preservation of Wilderness Areas: Hearings before the Subcommittee on Public Lands*, 92nd Cong., 2nd sess., May 5, 1972, 59–60.

10 Ibid., 500.

11 Senator Richard L. Neuberger, remarks on reintroduction of Wilderness Bill, *Congressional Record*, February 11, 1957, page 32 of reprint.

12 House Committee on Interior and Insular Affairs, *Designation of Wilderness Areas: Hearings before the Subcommittee on Public Lands and National Parks and Recreation*, 91st Cong., 2nd sess., June 26, 1970, 500.

13 Kenneth A. Reid, "Let Them Alone!," *Outdoor America* 5, no. 1 (November 1939): 6.

14 *Mount Naomi Wilderness Boundary Adjustment Act 2003*, Public Law 108–95, 117 Stat. 1165, October 3, 2003.

15 Robert Marshall to Harold Ickes, "Suggested Program for Preservation of Wilderness Areas," April 1934, National Archives, Record Group 79, File 601–12, Parks-General Lands-General-Wilderness Areas, pt. 1.

Chapter 2: Wilderness: From Concept to Action

1 Aldo Leopold, "The Last Stand of the Wilderness," *American Forests and Forest Life* 31, no. 382 (October 1925): 602.

2 "The Wilderness Society Platform," *The Living Wilderness* 1, no. 1 (September 1935): 2, reprinting a statement drafted that January.

3 Roderick F. Nash, *Wilderness and the American Mind*, 4th ed. (New Haven, Conn.: Yale University Press, 2001), xi and xiii.

4 Ibid., quoting Standing Bear, *Land of the Spotted Eagle* (Boston: Houghton-Mifflin Company, 1933), 38.

5 As quoted by Nash, *Wilderness and the American Mind*, 67.

6 Stephen J. Pyne, *How the Canyon Became Grand: A Short History* (New York: Viking Penguin, 1998), 2–3.

7 Ibid., 53.

8 John Muir, *The Yosemite* (New York: The Century Company, 1912), 256.

9 Quoted in Linne Marsh Wolfe, ed., *John of the Mountains: The Unpublished Journals of John Muir* (Boston: Houghton, Mifflin and Company, 1938), 317.

10 John Muir, "The American Forests," chapter 10, *Our National Parks*, (Boston: Houghton, Mifflin and Company, 1901).

11 Aldo Leopold, "Why the Wilderness Society?" *The Living Wilderness* 1, no. 1 (September 1935): 6.

12 Robert Marshall, "The Universe of the Wilderness Is Vanishing," *Nature Magazine* 9, no. 4 (April 1937): 236.

13 Aldo Leopold, *A Sand County Almanac* (New York: Oxford University Press, 1949), 167.

14 Aldo Leopold, "The Wilderness and Its Place in Forest Recreational Policy," *Journal of Forestry* 19,

no. 7 (November 1921): 720.

15 Dr. Charles C. Adams, "Ecological Conditions in National Forests and in National Parks," *Scientific Monthly* 20, no. 6 (June 1925): 562.

16 Letter, Louis W. Waldorf (Vice President for Oregon, FWOC) to Frank A. Kittridge (Regional Director, NPS, San Francisco), December 27, 1939, National Archives, Record Group 79, File 601-13, pt. 1.

17 Horace M. Albright, "The Everlasting Wilderness," *Saturday Evening Post* 201, no. 13 (September 29, 1928): 28.

18 Robert Marshall to Harold L. Ickes, May 21, 1937, Wilderness Society files and author's files.

19 Leopold, "The Wilderness and Its Place," 718.

20 Aldo Leopold, "Wilderness as a Form of Land Use," *Journal of Land and Public Utility Economics* 1 (October 1925): 401.

21 Aldo Leopold, "Conserving the Covered Wagon," *Sunset* (March 1925): 56.

22 Report of the Forester (Washington, D.C.: Government Printing Office, 1926), 36; and the same report for 1927, 13, both as cited in Dennis M. Roth, *The Wilderness Movement and the National Forests* (College Station, Tex.: Intaglio Press, 1988), 3.

23 Regulation L-20, October 30, 1929, as quoted in Roth, *The Wilderness Movement*, 3.

24 Mimeographed supplement to the Forest Service Administrative Manual sent to Forest Service Districts, June 29, 1929, as quoted in James P. Gilligan, "The Development of Policy and Administration of Forest Service Primitive and Wilderness Areas in the Western States" vol. II, appendix A (Ph.D. dissertation, University of Michigan, 1953), 1–2.

25 Robert Marshall, "The Problem of the Wilderness," *Scientific Monthly* (February 1930): 145.

26 Ibid., 148.

27 Robert Marshall to Harold Ickes, February 27, 1934, National Archives, Record Group 79, File 601-12, Parks-General-Lands-General-Wilderness Areas, pt. 1, 3.

28 Robert Marshall to Harold Ickes, "Suggested Program for Preservation of Wilderness Areas," April 1934, National Archives, Record Group 79, File 601-12, Parks-General-Lands-General-Wilderness Areas, pt. 1, 8.

29 Marshall, "The Problem of the Wilderness,"148.

30 H. H. Chapman, "National Parks, National Forests and Wilderness Areas," *Journal of Forestry* 36, no. 5 (May 1938): 474.

31 Robert Marshall, Memorandum for the Secretary, May 25, 1934, copy in Wilderness Society files and author's files.

32 Robert Marshall, "Statement on Behalf of the Wilderness Society," House Committee on Public Lands, *Mount Olympus National Park: Hearings before the House Committee on Public Lands*, 74th Cong., 2nd sess., April 23, 24, 25, 27, 28, 29, 30, May 1 and 5, 1936, 276; and Minutes, Fifth Annual Meeting of the Council, The Wilderness Society, March 25, 1939, 2, copies in Wilderness Society files and author's files.

33 Representative Bertrand W. Gearhart, testimony in House Committee on Public Lands, *John Muir-Kings Canyon National Park: Hearings before the House Committee on Public Lands on H.R. 3794*, 76th Cong., 1st sess., March–April 1939, 163.

34 U.S Department of the Interior, Memorandum for the Press, January 3, 1939, National Archives, Record Group 48, NPS 1933–42.

35 H.R. 3648, 76th Cong., 1st sess., introduced February 2, 1939, and S. 1188, introduced February 6, 1939. (Undated draft letter, Secretary of Agriculture to Vice-President, U.S. Senate; and Memorandum for Files, C.M. Granger, Acting Chief, Forest Service, June 30, 1939, both in

National Archives, Record Group 95, U, Legislation, Federal, S. 1188.)

36 Robert Marshall, Memorandum for Mr. Granger [Assistant Chief C. W. Granger], June 12, 1939, attaching two earlier memos on the same point, National Archives, Record Group 95, U, Legislation, Kings Canyon National Park, copy in author's files.

Chapter 3: Wilderness: "There Ought to Be a Law"

1 James M. Glover, *A Wilderness Original: The Life of Bob Marshall* (Seattle: The Mountaineers, 1986), 271. His younger brother, George, emphatically denied that Bob had any congenital heart problem.

2 Robert Sterling Yard to Bernard Frank, September 13, 1937, as cited in Roderick Nash, *Wilderness and the American Mind, 3rd ed.* (New Haven, Conn.: Yale University Press, 1982), 206.

3 Harvey Broome to Olaus Murie, August 1, 1940, Wilderness Society files and author's files.

4 James P. Gilligan, "The Contradiction of Wilderness Preservation in a Democracy," *Proceedings, Society of American Foresters,* Annual Meeting (October 24–27, 1954), 120–121.

5 Quoted in Kenneth A. Reid, "Let Them Alone!" *Outdoor America* 5, no. 1 (November 1939): 6.

6 John R. Bernard, "Forest Wilderness: How dedicated areas are being reclassified," *Sierra Club Bulletin* 41, no. 1 (January 1956): 26.

7 Reid, "Let Them Alone!"

8 Donald C. Swain, *Federal Conservation Policy 1921–1933, vol. 76, University of California Publications in History* (Berkeley: University of California Press, 1963), 130.

9 Harvey Broome, quoted in "1906 ... Howard Clinton Zahniser ... 1964," *The Living Wilderness* 28, no. 85 (Winter-Spring 1964): 3.

10 Olaus J. Murie, editorial, *The Living Wilderness* 10, no. 14 and 15 (December 1945): frontispiece.

11 H.R. 2142, 79th Cong., 1st sess.

12 Benton MacKaye to Representative Daniel K. Hoch, January 31, 1946, Wilderness Society files and author's files.

13 Benton MacKaye to Howard Zahniser, January 31, 1946, Wilderness Society files and author's files. A photo of Zahniser delivering MacKaye's draft bill to Hoch was printed in *The Living Wilderness* 11, no. 16 (March 1946): 5.

14 Robert Marshall to Harold Ickes, "Suggested Program for Preservation of Wilderness Areas," April 1934, National Archives, Record Group 79, File 601-12, Parks-General-Lands-General-Wilderness Areas, pt. 1, 8.

15 David Brower, "Wilderness and the Constant Advocate," *Sierra Club Bulletin* 49, no. 6 (September 1964): 3.

16 C. Frank Keyser, Committee on Merchant Marine and Fisheries, The Preservation of Wilderness Areas: An Analysis of Opinion on the Problem by *C. Frank Keyser, Regional Economist, Legislative Reference Service, Library of Congress,* August 24, 1949, Committee Print No. 19, 10.

17 Ibid., 65–66.

18 Zahniser's extensive submission to the Library of Congress study provides a window into his thinking at that time. See Memorandum for the Legislative Reference Service, March 1, 1949, in Senate Committee on Interior and Insular Affairs, *National Wilderness Preservation Act: Hearings before the Committee on Interior and Insular Affairs on S. 1176,* 85th Cong., 1st sess., June 19 and 20, 1957, 165–197.

19 Howard Zahniser, "How Much Wilderness Can We Afford to Lose?" in *Wildlands in Our Civilization* (San Francisco: Sierra Club, 1964), 51. This address was also printed in the *Sierra Club Bulletin* (April 1951).

20 Ibid., 50.

21 John N. Spencer to Richard M. Leonard, November 30, 1950, as quoted in Nash, *Wilderness and the American Mind, 4th ed.*, 212.

22 Minutes of the Annual Meeting of the Council, The Wilderness Society, September 3–7, 1955, 11, Wilderness Society files and author's files.

23 Quote attributed to Representative Wayne Aspinall, May 1954, included in a pamphlet, *What Is Your Stake in Dinosaur?* (San Francisco: Trustees for Conservation, 1954), author's files.

24 Mark W. T. Harvey, *A Symbol of Wilderness: Echo Park and the American Conservation Movement* (Albuquerque: University of New Mexico Press, 1994), 5.

25 One "price" of the Echo Park victory, in a fight deliberately defined as about protecting the integrity of the then-existing units of the National Park System, was the inclusion of authorization for the Glen Canyon dam in the Colorado River Storage Project Act from which the Echo Park dam had been removed. Glen Canyon was not then a unit of the National Park System; its loss, contrary to a common misunderstanding, was not the result of a trade-off or compromise by the conservationists to gain defeat of the Echo Park dam. Conservation leaders became aware of the awesome wild values of Glen Canyon, which were little known in the early 1950s, far too late. By that time attempting to change the firmly established definition and focus of the congressional fight was not politically feasible. Protection for Glen Canyon had not been discussed or embraced by the broader conservation coalition or its congressional champions as part of the campaign they had mobilized to defeat the Echo Park dam. They had, years earlier, chosen to focus on the question of National Park System integrity; changing the focus so much later in the game would likely have split their coalition and alienated some of their congressional support. This is not an excuse, simply the historical reality.

   The drowning of Glen Canyon—"the place no one knew"—was a tragic loss of a magical canyon wilderness. It became a potent symbol for the wilderness movement. See Harvey, *A Symbol of Wilderness*, especially pages 55–56, and David Brower, footnote to Clinton P. Anderson, "Changing Public Opinion—As a Legislator Sees It," *Sierra Club Bulletin* 49, no. 9, (December 1964): 42.

26 Howard Zahniser, "Wilderness Forever" (address to the Seventh Biennial Wilderness Conference, Sierra Club, San Francisco, April 8, 1961), 5. Typescript in Wilderness Society files and author's files.

27 Charlotte E. Mauk, "Groundwork for Cooperation: The Four Wilderness Conferences," *Sierra Club Bulletin* 41, no. 1 (January 1956): 18.

28 "The Need for Wilderness Areas," extension of remarks of Honorable Hubert H. Humphrey, *Congressional Record*, June 1, 1955, including the full text of Zahniser's speech. Quoted from a booklet reprint in author's files.

29 Howard Zahniser to George Marshall, April 3, 1956, Wilderness Society files and author's files.

30 George Marshall to Howard Zahniser, April 11, 1956, 2, Wilderness Society files and author's files.

31 Richard E. McArdle, statement, Senate Committee on Interior and Insular Affairs, *National Wilderness Preservation Act: Hearings before the Committee on Interior and Insular Affairs on S. 1176*, 85th Cong., 1st sess., June 19–20, 1957, 93.

32 James P. Gilligan, "The Contradiction of Wilderness Preservation in a Democracy," *Proceedings, Society of American Foresters*, Annual Meeting (October 24–27, 1954), 120.

33 Ibid., 121.

34 Senator Richard L. Neuberger, speaking on Proposed National Wilderness Preservation System, *Congressional Record*, 85th Cong., 1st sess., February 11, 1957, 103, pt. 2: 1906.

35 Trustees for Conservation, pamphlet, San Francisco, June 5, 1956. Wilderness Society files.

36 Senator Gordon Allott, Senate Committee on Interior and Insular Affairs, *Preservation of Wilderness*

Areas, Hearing before the Subcommittee on Public Lands on S. 2453 and Related Wilderness Bills, 92nd Cong., 2nd sess., May 5, 1972, 64.

37 J. Michael McCloskey, "The Wilderness Act of 1964: Its Background and Meaning," *Oregon Law Review* 45 (June 1966): 298.

38 Senator Robert C. Byrd, *Congressional Record*, September 6, 1961, 17241.

39 Representative John P. Saylor, Minority Views, in *Providing for the Preservation of Wilderness Areas, for the Management of Public Lands, and for other Purposes*, House Report 87-2521, October 3, 1962, 118 and 138.

40 This letter was not published by the *Times* due to a newspaper strike, but, with their permission, it was printed in the *Sierra Club Bulletin* (January 1959) and then reprinted in *The Living Wilderness* (Autumn 1958) and inserted in the *Congressional Record* by Senator Humphrey, February 19, 1959, as he reintroduced the Wilderness Bill in refined form in the new Congress.

41 Representative Wayne Aspinall, *Congressional Record*, September 20, 1962, 20202.

42 The Senate vote margins both years exaggerate the opposing vote. Because they were holding out for major changes they had been unable to pass as amendments in the Senate, a number of westerners chose to vote against final passage. This was intended to keep their issues alive in the House debate by weakening the perception of overwhelming Senate approval. After the compromises in 1964 met many of their requirements, the Senate ultimately passed the final version of the Wilderness Act in a unanimous voice vote.

43 This episode is thoroughly covered in Steven C. Schulte, *Wayne Aspinall and the Shaping of the American West* (Boulder: University Press of Colorado, 2002). For his master's thesis my longtime colleague Jack Hession of the Sierra Club interviewed Chairman Aspinall about his conversation with President Kennedy. See Jack M. Hession, "The Legislative History of the Wilderness Act" (Master's thesis, San Diego State College, 1967): 207–208. See also "JFK Lauded Aspinall for Wilderness Bill," *Denver Post*, December 1, 1963.

44 James L. Sundquist, *Politics and Policy: The Eisenhower, Kennedy, and Johnson Years* (Washington, D.C.: The Brookings Institution, 1968): 360–361.

45 I am often asked who cast the lone vote against the Wilderness Act. It was Representative Joe Pool, a Texas Democrat who was a freshman representative in 1964 and who served until his death in 1968. He was not a member of the congressional committee involved and did not participate in the debate over the bill. I have not found any explanation for his vote.

46 David Brower, "Wilderness and the Constant Advocate," *Sierra Club Bulletin* 49, no. 6 (September 1964): 3.

47 Olaus Murie, Report of the Director to the Council of The Wilderness Society, June 18, 1950, in Benton MacKaye, "The Call," June 20, 1950, 5, issued for the 1950 meeting of the council, in Wilderness Society files and author's files.

48 Fred M. Packard to Howard Zahniser, April 13, 1951, Wilderness Society files and author's files.

49 Howard Zahniser, "The People and Wilderness," *The Living Wilderness* 28, no. 86, (Spring-Summer 1964): 41.

50 Howard Zahniser to Wayne Aspinall, July 19, 1963, Box 139, Wayne Aspinall Papers, Special Collections and Archives, Penrose Library, University of Denver. Copy in Wilderness Society files and author's files.

51 Howard Zahniser to Harvey Broome, May 3, 1963, Wilderness Society files and author's files.

Chapter 4: Putting the New Wilderness Act to Work

1 The other great defeat, as it was seen at the time, was the inclusion in the act of a nineteen-year

period during which new mining claims could still be staked within national forest wilderness areas. This was a source of great anxiety among wilderness advocates at the time I came into the movement in 1967. In practice, however, there is no evidence that any real harm resulted. While some new claims certainly were staked, to my knowledge there has not been any real threat of actual mining on any such claims made during that nineteen-year period which closed on December 31, 1983.

2 Senator Henry M. Jackson to a constituent, March 15, 1963, Senate Interior and Insular Affairs Committee, S. 2 S.4 (cont.), National Archives, Record Group 46, as cited in Dennis M. Roth, *The Wilderness Movement and the National Forests* (College Station, Tex.: Intaglio Press, 1988), 9–10.

3 Stewart M. Brandborg, "Executive Director's Report to Council 1966–67," Wilderness Society Papers, Western History Department, Denver Public Library, Box 2, as cited in Roth, *The Wilderness Movement*, 13–14.

4 Senator Frank Church, dialogue with Senator Gordon Allott, Senate Committee on Interior and Insular Affairs, *Preservation of Wilderness Areas: Hearing before the Subcommittee on Public Land, on S. 2453 and Related Wilderness Bills*, 92nd Cong., 2nd sess., May 5, 1972, 65.

5 Quoted in Ronald G. Strickland, "Ten Years of Congressional Review Under the Wilderness Act of 1964: Wilderness Classification Through Affirmative Action," (unpublished dissertation, Georgetown University, 1976), 56, as cited by Roth, *The Wilderness Movement*, 21.

6 "Area History," text on official map, "Shining Rock Wilderness and Middle Prong Wilderness, Pisgah National Forest, North Carolina," Forest Service Recreational Guide R8-RG 23, revised June 1993.

7 "New Wild Area Established in North Carolina," U.S. Forest Service press release, May 13, 1964, Atlanta.

8 "Shining Rock Wild Area—North Carolina: Proposal," Southern Region, U.S. Forest Service, 3. This report is undated, but the Forest Service press release announcing designation of the wild area noted that "The plans for making this a wild area were released in September 1963."

9 S. 3699 (92nd Congress, 2nd sess.). Statement, Senator George Aiken (R-VT), *Congressional Record*, June 13, 1972, 20570.

10 John P. Saylor, "Legislation to Save Eastern Wilderness," *Congressional Record*, January 11, 1973, 849.

11 Ibid.

12 Ibid.

13 *Congressional Record*, January 11, 1973, 754.

14 Ibid., 757.

15 Senate Committee on Interior and Insular Affairs, *Eastern Wilderness Areas: Hearing before the Subcommittee on Public Lands on S. 316*, 93rd Cong., 1st sess., February 21,1973, 31.

16 Ibid., 32.

17 Ibid., 33.

18 Aldo Leopold, *A Sand County Almanac* (New York: Oxford University Press, 1949), 189.

19 Section 101(a)(5), Vermont Wilderness Act of 1984, P.L. 98–322, June 19, 1984, 98 Stat. 253.

20 Senator Philip A. Hart, in Senate Committee on Interior and Insular Affairs, *Preservation of Wilderness Areas: Hearing before the Subcommittee on Public Lands on S. 2453 and Related Wilderness Bills*, 92nd Cong., 2nd sess., May 5, 1972, 50.

21 Senator Frank Church, in Senate Committee on Interior and Insular Affairs *Preservation of Wilderness Areas: Hearing before the Subcommittee on Public Lands on S. 2453 and Related Wilderness Bills*, 92nd Cong., 2nd sess., May 5, 1972, 59–60.

22 Lois Crisler, *Arctic Wild* (New York: The Lyons Press, 1999), 38.

23 Ibid, 196.

24 Representative John Dingell, debate on H.R. 19007, omnibus mandate areas bill, *Congressional Record*, September 21, 1970, 32754.

Chapter 5: Expanding the Scope of Wilderness Preservation

1 Letter from fourteen Republican senators to Secretary of the Interior Gale Norton, April 9, 2003, 2.

2 Ibid.

3 *Congressional Record*, July 30, 1964, 17430.

4 David R. Brower, "De Facto Wilderness: What Is Its Place?" *Wildlands in Our Civilization*, (San Francisco: Sierra Club, 1964): 103 and 109.

5 Representative John P. Saylor, "The New Wilderness Bill," *Congressional Record*, November 7, 1963, cited from a booklet reprint, 2.

6 Stewart M. Brandborg, "Acting Executive Director's Report to the Council, 1963–64," 7.

7 Neil Rahm, speech, March 13, 1971, in records of Robert Morgan, as quoted in Dennis M. Roth, *The Wilderness Movement and the National Forests* (College Station, Tex.: Intaglio Press, 1988), 35.

8 *Congressional Record*, October 14, 1970, 36758.

9 U.S. House of Representatives, *Designating Certain Endangered Public Lands for Preservation as Wilderness, Providing for the Study of Additional Endangered Public Lands for such Designation, Furthering the Purposes of the Wilderness Act of 1964, and for other Purposes*, House Report No. 95–540, July 27, 1977, 5.

10 Robert Sterling Yard, "For a Wilderness Policy: To Protect from Unappreciative Encroachment Rare Primitive Areas Which Still Remain," *National Parks Bulletin* 7, no. 47 (January 1926): 8.

11 Howard Zahniser, "A Statement on the Wilderness Act," presented at Senate Committee on Interior and Insular Affairs, *Hearing before the Subcommittee on Public Lands on the Wilderness Act*, May 7–9, 1962, 15. Typescript in Wilderness Society files and author's files.

12 Aldo Leopold, "Why the Wilderness Society?" *The Living Wilderness* 1, no. 1 (September 1935): 6.

13 Aldo Leopold, "The Last Stand of the Wilderness," *American Forests and Forest Life* 31, no. 382 (October 1925): 603.

14 The Roadless Area Conservation Rule has been codified as 36 CFR 294, Subpart B—Protection of Inventoried Roadless Areas, but is subject to possible repeal or revision as described here.

15 Federal Register 37240 (June 28, 2004), Sequence No. 209, State Petitions for Inventoried Roadless Area Management.

16 The definitive guide to the complex details of the mammoth Alaska National Interest Lands Conservation Act is *Alaska National Interest Lands Conservation Act: Citizens' Guide* (Washington, D.C.: The Wilderness Society, undated).

17 Tongass Timber Reform Act, November 20, 1990, Public Law 101-626, 104 Stat. 4426.

18 T. H. Watkins, "Alaska and the Weight of History," *Celebrating Wild Alaska: Twenty Years of the Alaska Lands Act*, (Natural Resources Defense Council and Alaska Wilderness League, 2000), frontispiece (reprint of a 1990 essay).

Chapter 6: Wilderness Yet to Save

1 David R. Brower, "De Facto Wilderness: What Is Its Place?" *Wildlands in Our Civilization*, (San Francisco: Sierra Club, 1964): 103.

2 Aldo Leopold to Bob Marshall, July 10, 1937, author's files.

3 Robert Marshall, "A Plea for the Old Wilderness," *New York Times Magazine* (April 25, 1937).

4 The 319 million-acre figure must be taken as very rough and preliminary. It deals with *only* roadless

"unprotected wilderness" lands on national forests and public lands administered by the U.S. Forest Service and the Bureau of Land Management. A computerized mapping analysis identified these unprotected wilderness lands as not having roads (down to a minimum size of 1,000 acres), but took no other factors into account. The analysis did *not* cover factors best assessed by on-the-ground inventory. However, citizen groups have conducted their own intensive inventories of many of these roadless lands to refine proposals for wilderness designation, and this work continues.

5 David Brower, "De Facto Wilderness: What Is Its Place?" 108.

6 Quoting standard language from 1980s statewide wilderness bills; this is from section 5(b)(2)(b) of the 1989 Nevada Wilderness Protection Act, Public Law 101–195, December 5, 1989. This requirement is part of the release language negotiated in the aftermath of a State of California lawsuit and is found in laws affecting most states with national forest lands.

7 "Record of Decision for the Final Environmental Impact Statement and Revised Land and Resource Management Plan (Revised Forest Plan), Wasatch-Cache National Forest," signed by Jack G. Troyer, Regional Forester, Intermountain Region, dated March 19, 2003, 5, 8.

8 Ibid., 27–28.

9 Stipulation and Joint Motion to Enter Order Approving Settlement and to Dismiss the Third Amended and Supplemented Complaint, *State of Utah v. Norton*, U.S. District Court, District of Utah, Central Division, typescript filed April 11, 2003, and approved by the federal judge April 24, 2003: 14.

10 The Utah state director of the Bureau of Land Management ordered employees to make issuing new drilling permits their highest priority, noting that wilderness reviews "hinder, rather than promote access to oil and gas resources on public lands." Staff were directed to focus their limited time on processing drilling applications. Bureau of Land Management, Information Bulletin No. UT 2002-008, January 4, 2002.

11 Proposed bills S. 2342 and H.R. 4202 (108th Congress, 2nd sess.).

12 "Wilderness Mountain Bicycling Groups Reach Agreement on Jefferson National Forest Wilderness Proposal," International Mountain Biking Association press release, April 2004. www.imba.com/resources/wilderness/Virginia/Jefferson_Wild_pr.html, accessed 7/6/04.

13 Identical wording in both ads; quoted from version on the Washington State Snowmobile Association Web site: http://www.wssa.us/Wilderness_AD_04.pdf, accessed 5/03/04.

14 Clark L. Collins (executive director, BlueRibbon Coalition) and Rod Jones (Northwest Motorcycle Association), "No More Wilderness," apparently from a back issue of *BlueRibbon Magazine*, archived without source at www.sharetrails.org/index.cfm?page=490, accessed 7/5/04.

15 Quoting "Backcountry Designation: Legislative Objectives," on BlueRibbon Coalition Web site: www.sharetrails.org.index.cfm?page=39; accessed 5/3/04.

16 James P. Gilligan, "The Contradiction of Wilderness Preservation in a Democracy," *Proceedings, Society of American Foresters*, Annual Meeting (October 24–27, 1954), 122.

17 Bill Dart, "The National Public Lands Report," *BlueRibbon Magazine* (issue date not provided but by context early 2004), www.sharetrails.org/magazine.cfm?story=376, accessed 5/3/04. Dart's implication — commonly made by extractive industry and off-road vehicle groups — that "multiple use land" is distinguished from statutory wilderness is flat wrong, as Congress took care to spell out in the words of the Multiple-Use Sustained-Yield Act of 1960 and its legislative history, Public Law 86-517, June 12, 1960, 74 Stat. 215. Section 2 of that Act, as codified at 16 U.S.C. 529, makes this clear:

Sec. 529. - Authorization of development and administration consideration to relative values of resources; areas of wilderness

The Secretary of Agriculture is authorized and directed to develop and administer the renewable surface resources of the national forests for multiple use and sustained yield of the several products and services obtained therefrom. In the administration of the national forests due consideration shall be given to the relative values of the various resources in particular areas. The establishment and maintenance of areas of wilderness are consistent with the purposes and provisions of sections 528 to 531 of this title.

18 Clark Collins, "Wilderness Is Not Good for Recreation" (date not provided), www.sharetrails.org/index.cfm?page=489, accessed 5/3/04.

19 Collins and Jones, "No More Wilderness."

20 36 CFR 219.6(a)(2).

21 H. Ken Cordell, Michael A. Tarrant, and Gary T. Green, "Is the Public Viewpoint of Wilderness Shifting?" *International Journal of Wilderness* 9, no. 2 (August 2003): 31.

22 Paul Arndt, Natural Resource Planner, Southern Regional Office, U.S. Forest Service, preface to *Public Survey Report: Public Use and Preferred Objectives for Southern Appalachian National Forests*, (U.S. Forest Service and University of Tennessee [a collaboration of the Southern Regional Office, USFS; Southern Research Station, USFS; and Human Dimensions Research Lab, Department of Forestry, Wildlife, and Fisheries, University of Tennessee], July 2002), iii.

23 This and the data in the following paragraph are drawn from *Public Survey Report: Public Use and Preferred Objectives for Southern Appalachian National Forests* (U.S. Forest Service and University of Tennessee [a collaboration of the Southern Regional Office, USFS; Southern Research Station, USFS; and Human Dimensions Research Lab, Department of Forestry, Wildlife, and Fisheries, University of Tennessee], July 2002), 15, 34–35.

24 *Public Survey Report: Public Use and Preferred Objectives for Southern Appalachian National Forests*, 34 and 36.

25 Ibid., 41.

26 Ibid.

27 For a compilation of public opinion polls concerning wilderness taken by diverse polling organizations between 1999 and 2002, see *A Mandate to Protect America's Wilderness: A Comprehensive Review of Recent Public Opinion Research*, Campaign for America's Wilderness, January 2003, available at http://leaveitwild.org/reports/reports.html.

Chapter 7: Wilderness Politics

1 Robert T. Dennis, Izaak Walton League of America, "The Wilderness Act," paper given at the University of Michigan, November 5, 1964, 1. Typescript in author's files.

2 David Brower, "Wilderness and the Constant Advocate," reprinted in *The Living Wilderness* (Spring-Summer 1964): 42.

3 Harvey Broome, quoted in "1906 ... Howard Clinton Zahniser ... 1964," *The Living Wilderness* 28, no. 85 (Winter-Spring 1964): 3.

4 Editorial, "This Takes Time," *Daily Astorian*, March 19, 2004.

5 Lisa Riley Roche, "Ideas for Wilds Run Gamut at U.," *Deseret Morning News*, April 17, 2004.

6 Robert Marshall, "The Universe of the Wilderness Is Vanishing," *Nature Magazine* (April 1937): 238.

7 Roche, "Ideas for Wilds Run Gamut at U."

8 Aldo Leopold, "The Last Stand of the Wilderness," *American Forests and Forest Life* 31, no. 382 (October 1925): 601.

9 Organizing committee, "Reasons for a Wilderness Society," paper issued January 21, 1935.

10 From special run of NSRE data prepared for author by Gary Green, February 17, 2003.

11 Dan Wood, Director of Government Relations, Washington State Farm Bureau, to Chairman Richard Pombo, Committee on Resources, U.S. House of Representatives, April 20, 2004.

12 "The Great 'Wild Sky' Deception!!!," flyer from Wild Sky event, 2003.

13 House Committee on Interior and Insular Affairs, *Designation of Wilderness Areas: Hearings before the Subcommittee on Public Lands and National Parks and Recreation*, 91st Cong., 2nd sess., June 26, 1970, 501.

14 "OHV Talking Points," in "Our OHV Future is Now" by Land-Use Lump, *Dirt Bike Magazine*, November 19, 2002, www.dirtbikemagazine.com/archives.asp?vaR=READ&ID=318.

15 Rainer Huck, "Wilderness Cuts Off Multiple-Use Access to Public Lands," op-ed, *Salt Lake Tribune*, June 29, 2003.

16 http://www.forestsforpeople.com//, accessed 7/5/04.

17 *A Mandate to Protect America's Wilderness: A Comprehensive Review of Recent Public Opinion Research*, Campaign for America's Wilderness, January 2003, available at http://leaveitwild.org/reports/reports.html.

18 NSRE, 2000–2001, 10,382 respondents nationwide. Question: "How do you feel about designating more of the federal lands in your state as wilderness? Would you say you strongly favor, somewhat favor, neither favor or oppose, somewhat oppose, or strongly oppose this idea?"

19 NSRE, 2000–2001, 1,539 self-identified Hispanic respondents nationwide.

20 The NSRE survey research scientists pretested the wording of this introductory statement. They found most people do not relate to a number like "106 million acres," the current size of the wilderness system, so referred to the number of separately named wilderness areas (note that this number is out of date). Moreover, their testing found that including or not including this descriptor made no difference in responses.

21 H. Ken Cordell, Michael A. Tarrant, and Gary T. Green, "Is the Public Viewpoint of Wilderness Shifting?" *International Journal of Wilderness* 9, no. 2 (August 2003): 30–31.

22 Robert Pear, "One Step at a Time: Clinton Seeks More Help for Poor and Elderly," *New York Times*, February 8, 2000.

23 Katharine Q. Seelye, "Industry Seeking Rewards from G.O.P.-Led Congress," *New York Times*, December 3, 2002, A28.

24 Roger Scruton, "A Question of Temperament," *Wall Street Journal*, op-ed, December 3, 2002.

25 Theodore Roosevelt's speech at Grand Canyon, May 6, 1903, text from *Coconino Sun*, Flagstaff, Arizona, May 9, 1903, 1.

26 Senator John McCain, "Nature Is Not a Liberal Plot," *New York Times*, November 22, 1996.

27 Bertram M. Gross, *The Legislative Struggle: A Study in Social Combat* (New York: McGraw-Hill Book Company, 1953; reprinted by Greenwood Press, Westport, Conn., 1978), 96.

28 Robert A. Caro, *The Years of Lyndon Johnson: Master of the Senate* (New York: Alfred A. Knopf, 2002), 460; quoting from Hubert H. Humphrey, *The Education of a Public Man: My Life and Politics* (Garden City, N.Y.: Doubleday, 1976): 136, 137.

29 Gross, *The Legislative Struggle*, 227.

30 Humphrey quoted in Caro, *Master of the Senate*, 460.

31 Howard Zahniser to William R. Halliday, October 14, 1959, Wilderness Society files and author's files.

32 Howard Zahniser to Harvey Broome, May 3, 1963, Wilderness Society files and author's files.

33 Howard Zahniser, "The People and Wilderness," *The Living Wilderness* 28, no. 86 (Spring-Summer 1964): 39. This issue of the magazine was not issued until after Zahniser's death in May 1964.

34 Ibid., 41.

35 Ibid.

36 American Viewpoint, Hood River County, Oregon, poll conducted April 6–7, 2004, among 350 respondents.

37 Moore Information Public Opinion Research, "Idaho Voters & Management of the Boulder-White Cloud Mountains and the Owyhee-Bruneau Canyonlands," May 1, 2004, 1.

38 W. O. Douglas, foreword to *The Wild Cascades: Forgotten Parkland*, by Harvey Manning (San Francisco: Sierra Club, 1965): 14–15.

Chapter 8: The Challenges of Wilderness Stewardship

1 Vance G. Martin, publisher's preface to *Wilderness Management: Stewardship and Protection of Resources and Values*, 3rd ed., by John C. Hendee and Chad P. Dawson (Golden, Colo.: The Wild Foundation and Fulcrum Publishing, 2002), ix.

2 Howard Zahniser, "Guardians Not Gardeners," editorial, *The Living Wilderness* (Spring-Summer 1963): 2.

3 Ibid.

4 David Brower, "De Facto Wilderness: What is Its Place?," *Wildlands in Our Civilization*, Sierra Club, 1964, 103.

5 Zahniser, "Guardians Not Gardeners."

6 Minnesota Public Interest Research Group v. Butz, Memorandum and Order, U.S. District Court for the District of Minnesota, August 13, 1975, 401 F. Supp. 1276, esp. 1331.

7 16 U.S.C. 1132(c).

8 "Supplementary Statement of Howard Zahniser, Executive Director of the Wilderness Society," in Senate Committee on Interior and Insular Affairs, *National Wilderness Preservation Act: Hearings before the Committee on Interior and Insular Affairs on S. 4*, 88th Cong., 1st sess., February 28 and March 1, 1963, 68.

9 Statement of Senator Clinton P. Anderson, in Senate Committee on Interior and Insular Affairs, *Wilderness Act: Hearings before the Senate Committee on Interior and Insular Affairs on S. 174*, 87th Cong., 1st sess., February 27–28, 1961, 2.

10 Parker v. United States, 309 F. Supp. 593, U.S. District Court for the District of Colorado, Memorandum Opinion and Order, February 27, 1970.

11 M.M. Nelson, Deputy Chief, U.S. Forest Service, Letter to Senator Clifford P. Hansen, February 15, 1968, in *Senate Committee on Interior and Insular Affairs San Gabriel, Washakie, and Mount Jefferson Wilderness Areas: Hearings before the Subcommittee on Public Lands*, Committee on Interior and Insular Affairs, 90th Congress, 2nd sess., February 19–20, 1968, 24.

12 Senator Frank Church, "The Wilderness Act Applies to the East," *Congressional Record*, January 16, 1973, 1251.

13 I am preparing a series of Briefing Papers focused on each of these nonconforming uses, as well as other issues in wilderness designation and stewardship. These are available at http://leaveitwild.org/reports/briefing.html. Additional papers, as well as updates on those already there, are added frequently.

14 Robert Marshall, "The Problem of the Wilderness," *Scientific Monthly*, February 1930: 146.

15 Howard Zahniser, "A Statement on Wilderness Preservation in Reply to a Questionnaire" [submitted to the Legislative Reference Service, Library of Congress], March 1, 1949, reprinted in *National Wilderness Preservation Act*, Hearings before the Committee on Interior and Insular Affairs, U.S. Senate (85th Congress, 1st session), June 19-20, 1957: 192.

16 Senator Frank Church, during Senate floor debate on the Wilderness Bill, *Congressional Record*,

September 5, 1961, page 19622 (daily edition), emphasis added.

17 Arthur Carhart was a young Forest Service landscape architect in the early 1920s who was in contact with Aldo Leopold and whose own work subsequently helped implement early wilderness protection.

18 Senate Committee on Interior and Insular Affairs *Preservation of Wilderness Areas: Hearing before the Subcommittee on Public Lands on S. 2453 and Related Wilderness Bills*, 92nd Cong., 2nd sess., May 5, 1972, 61–62.

19 *Ensuring the Stewardship of the National Wilderness Preservation System: A Report to the USDA Forest Service, Bureau of Land Management, US Fish and Wildlife Service, National Park Service, US Geological Survey*, Pinchot Institute for Conservation, September 2001, i, at www.pinchot.org/pic/ wilderness_report.pdf.

20 Ibid., iii.

21 Section 9, Washington State Wilderness Act of 1984, Public Law 98–339; 98 Stat. 30.

22 *Designating Certain National Forest System Lands in the State of Utah for Inclusion in the National Wilderness Preservation System to Release Other Forest Lands for Multiple Use Management, and for other Purposes*, House Report 98–1019, pt. 1, September 13, 1984, 4–5.

23 John C. Hendee and Chad P. Dawson, *Wilderness Management: Stewardship and Protection of Resources and Values*, 3rd ed. (Golden, Colo.: The Wild Foundation and Fulcrum Publishing, 2002).

24 Lyle F. Watts, "Can We Save Our Wilderness Areas in the National Forests?" talk at the National Citizens Conference on Planning for City, State, and Nation, Washington, D.C., May 12–17, 1950. Reprinted in *The Living Wilderness* (Spring 1950).

Chapter 9: Wilderness in Other Realms

1 Cheyenne River Sioux Tribe, Resolution No. 210-02-CR, June 5, 2003.

2 Tribal Ordinance 79A, 1982, 1.

3 Chad P. Dawson and Pauline Thorndike, "State-Designated Wilderness Programs in the United States," *International Journal of Wilderness*, 8, no. 3 (December 2002). 21–26.

4 California Public Resources Code, Section 5093.33(a).

5 Article XIV of the New York State Constitution. Available at http://assembly.state.ny.us/leg/?co=16.

6 Ed Zahniser, *Where Wilderness Preservation Began: Adirondack Writings of Howard Zahniser* (Utica, N.Y.: North Country Books, 1992).

7 Vance G. Martin and Alan Watson, "International Wilderness," in *Wilderness Management: Stewardship and Protection of Resources and Values*, 3rd ed , by John C. Hendee and Chad P. Dawson (Golden, Colo.: The Wild Foundation and Fulcrum Publishing, 2002), 49.

8 Ibid.

9 Ibid., 50.

10 International Union for the Conservation of Nature, *Guidelines for Protected Area Management Categories*, Undated. These definitions, together with objectives for management, guidance for selection, and other details, are found in Part II, "The Management Categories," available on the worldwide web. This definition is at page 2. http://www.iucn.org/themes/wcpa/pubs/pdfs/ iucncategories.pdf, accessed 7/7/04.

11 Ibid., 3.

12 Martin and Watson, 94.

13 http://allafrica.com/stories/printable/200405060212.html.

Chapter 10: An Enduring Resource of Wilderness

1 Robert Marshall to Harold Ickes, "Suggested Program for Preservation of Wilderness Areas," April 1934, National Archives, Record Group 79, File 601–12, Parks-General Lands-General-Wilderness Areas, pt. 1.

2 Harvey Broome, quoted in "1906 ... Howard Clinton Zahniser ... 1964," *The Living Wilderness*, no. 86 (Winter-Spring 1964): 3.

3 Richard Bangs, "Second Mountain Syndrome," *NWA World Traveler*, Northwest Airlines (April 2004): 19.

4 Representative John Saylor, remarks before the Republican Delegations of the Eleven Western States, Eugene, Oregon, October 11, 1963, typescript in author's files.

5 David Brower, "Dave Brower's Thanks," *Sierra Club Bulletin* 41, no. 6 (July 1956): 7.

6 Henry David Thoreau, *Walden and Other Writings of Henry David Thoreau*, ed. Brooks Atkinson (New York: The Modern Library, 1950), 283.

7 Theodore Roosevelt, "Man in the Arena" speech, University of Paris, The Sorbonne, April 23, 1910.

8 Ted Monoson, Lee Washington Bureau, "Washington Politics, Montana Style," as printed in several Montana newspapers, including the Helena and Missoula papers, April 7, 2004.

9 Max Weber, "Politics as a Vocation," 'Politik als Beruf,' *Gesammelte Politische Schriften* (Muenchen, 1921), 396-450. Originally a speech at Munich University, 1918, published in 1919 by Duncker & Humblodt, Munich.

10 Aldo Leopold, "The Last Stand of the Wilderness," *American Forests and Forest Life* 31, no. 382 (October 1925): 602.

11 Frederick Jackson Turner, "Commencement Address, University of Washington, June 17, 1914," The Washington Historical Quarterly (October 1914). http://xroads.virginia.edu/~HYPER/TURNR/chapter11.html.

12 Terry Tempest Williams, *Red: Passion and Patience in the Desert* (New York: Pantheon, 2001), 215.

13 Howard Zahniser, "Address to 1961 Sierra Club Biennial Wilderness Conference," in *Voices for the Wilderness*, ed. William Schwartz (New York: Ballantine Books, 1969), 106.

Suggested Reading

History of Wilderness Ideas and Protection

The classic study of the intellectual evolution of the idea of wilderness and an overview of the policy history is Roderick Nash, *Wilderness and the American Mind* (New Haven, Conn.: Yale University Press, 1967). Professor Nash's magisterial study provides a rich bibliography. The fourth edition (2001) is available in paperback.

By focusing on a specific place, historian Stephen J. Pyne's *How the Canyon Became Grand: A Short History* (New York: Viking Penguin, 1998) brings the evolving appreciation of wildness alive. For the cultural history of one eastern wilderness (not yet designated by Congress) once heavily impacted by man, see Margaret Lynn Brown, *The Wild East: A Biography of the Great Smoky Mountains* (Gainesville: University Press of Florida, 2000).

In *John Muir and the Sierra Club: The American Conservation Movement* (Boston: Little Brown, 1981), Stephen Fox examines the history of the conservation and wilderness movements. A University of Wisconsin Press paperback is available.

For the history of national forest wilderness after passage of the Wilderness Act, with detail about the Forest Service's RARE-I and RARE-II roadless area efforts and how conservationists and Congress responded, see Dennis M. Roth, *The Wilderness Movement and the National Forests* (College Station, Tex.: Intaglio Press, 1988). Roth was the historian of the U.S. Forest Service.

A detailed history of nature preservation in the national parks and the tensions within the National Park Service over wilderness protection is found in Richard West Sellars's *Preserving Nature in the National Parks: A History* (New Haven, Conn.: Yale University Press, 1997). Students of wilderness protection in the national parks eagerly await publication of professor John Miles's exhaustive history of the National Park Service's love-hate relationship with the Wilderness Act, tentatively titled *Playground or Preserve: The Story of Wilderness in the National Parks*.

Histories of wilderness preservation on lands administered by the Fish and Wildlife Service and the Bureau of Land Management remain to be written, but the latter is touched upon in William L. Graf's broader study, *Wilderness Preservation and the Sagebrush Rebellions* (Savage, Md.: Rowman & Littlefield Publishers, 1990).

In *Driven Wild: How the Fight against Automobiles Launched the Modern Wilderness Movement* (Seattle: University of Washington Press, 2002) historian Paul Sutter teases out the fundamental forces in the transition from the philosophical phase of wilderness appreciation to the practical phase of finding ways to preserve wilderness areas.

James Morton Turner has recently completed an exhaustive study of the first three decades of implementing the Wilderness Act, covering all four wilderness agencies. His dissertation, *The Promise of Wilderness: A History of American Environmental Politics, 1964–1994* (Princeton, N.J.: Princeton University, 2004), is sure to reach book form in the near future.

Biography

Historian Mark W. T. Harvey's eagerly awaited biography on Howard Zahniser, to be published in 2005 or 2006, is *Guarding the Wilderness: The Conservation Career of Howard C. Zahniser* (Seattle: University of Washington Press, forthcoming).

Legislators:

After Howard Zahniser himself, no person had a greater impact on the success of the National Wilderness Preservation System than Representative Wayne N. Aspinall. The essential source is Stephen C. Schulte's *Wayne Aspinall and the Shaping of the American West* (Boulder, Colo.: University Press of Colorado, 2002), which captures in detail the relationship between Aspinall and Zahniser through the Echo Park and Wilderness Bill years.

There is yet no biography of Representative John P. Saylor. The life and career of the other original lead sponsor of the Wilderness Act, Senator Hubert H. Humphrey, are chronicled in numerous books. In 1961 leadership for the bill in the Senate shifted to Senator Clinton P. Anderson, whose work for the Wilderness Act is well told by Richard Allan Baker, head of the Senate Historical Office, in *Conservation Politics: The Senate Career of Clinton P. Anderson* (Albuquerque: University of New Mexico Press, 1985).

Implementing the Wilderness Act in the 1960s and 1970s was a major issue for three nationally prominent legislations. In *A Lifelong Affair: My Passion for People and Politics* (Washington, D.C.: The Francis Press, 2003), Bethine Church, Senator Church's widow, recounts their unique partnership in Idaho and national politics. *Mo: The Life and Times of Morris K. Udall* (Tucson: The University of Arizona Press, 2001) by Donald W. Carson and James W. Johnson sets the context of his leadership for wilderness, but also see Mo's wonderful memoir, *Too Funny to Be President* (New York: Henry Holt and Company, 1988), particularly the chapters on the Grand Canyon and Alaska. Wilderness figures prominently in the story of the extraordinary legislative dynamo Phillip Burton, captured so well by John Jacobs in *A Rage for Justice: The Passion and Politics of Phillip Burton* (Berkeley, Calif.: University of California Press, 1995).

Founders of The Wilderness Society:

Paul Sutter's *Driven Wild*, already mentioned, is in effect the first volume of a "biography" of The Wilderness Society and a collective biography of its founders.

There are three biographies of Aldo Leopold: Curt Meine, *Aldo Leopold: His Life and Work* (Madison: University of Wisconsin Press, 1991); Susan L. Flader, *Thinking Like a Mountain: Aldo Leopold and the Evolution of an Ecological Attitude Toward Deer, Wolves, and Forests* (Madison: University of Wisconsin Press, 1994); and Marybeth Lorbiecki, *Aldo Leopold: A Fierce Green Fire* (Helena and Billings, Mont.: Falcon Publishing, 1996).

For Bob Marshall's biography, see James M. Glover, *A Wilderness Original: The Life of Bob Marshall* (Seattle: The Mountaineers, 1986). In her dissertation, *Robert Marshall, Howard Zahniser, and the Wilderness Society: A Life History Study of Agency, Structure, and Wilderness Conservation in the United States* (University of Idaho, 2004), AnneMarie Lankard Moore traces the direct linkage between Marshall's thinking and the insights Howard Zahniser brought to his work conceiving the Wilderness Act.

The life and work of Benton MacKaye is elucidated in Larry Anderson's *Benton MacKaye: Conservationist, Planner, and Creator of the Appalachian Trail* (Baltimore, Md.: The John Hopkins University Press, 2002). Joe Paddock tells the story of Ernest Oberholtzer in *Keeper of the Wild: the Life of Ernest Oberholtzer* (St. Paul, Minn.: Minnesota Historical Society Press, 2001). Though not one of the founders, Sig Olson was a longtime leader of The Wilderness Society; his story is told by David Backes in *A Wilderness Within: The Life of Sigurd F. Olson* (Minneapolis: University of Minnesota Press, 1997).

Histories of Specific Wilderness Preservation Campaigns

The Echo Park dam fight was a central influence in the history of the wilderness movement. It is splendidly recounted in historian Mark W. T. Harvey's *A Symbol of Wilderness: Echo Park and the American Conservation Movement* (Albuquerque: University of New Mexico Press, 1994), available in paperback from the University of Washington Press.

In *Troubled Waters: The Fight for the Boundary Waters Canoe Area Wilderness* (St. Cloud, Minn.: North Star Press of St. Cloud, 1995), three participants—Kevin Proescholdt, Rip Rapson, and Miron L. Heinselman—focus on the campaign to pass the Boundary Waters Canoe Area Wilderness Act of 1978.

An excellent case study of designating wilderness on lands administered by the Bureau of Land Management is the history of the California Desert Protection Act of 1994 recounted in Frank Wheat, *California Desert Miracle: The Fight for Desert Parks and Wilderness* (San Diego: Sunbelt Publications, 1999).

Two books and a dissertation focus on the history of wilderness campaigns in the Pacific Northwest. The history of the Alpines Lakes Wilderness is told by a leading participant, David Knibb, in *Backyard Wilderness: The Alpine Lakes Story* (Seattle: The Mountaineers, 1982). William Ashworth chronicles the blocking of dams and protection of the wild and scenic river and wilderness in Hells Canyon in *Hells Canyon: The Deepest Gorge on Earth* (New York: Hawthorn Books, 1977). Kevin R. Marsh's dissertation, *Drawing Lines in the Woods: Debating Wilderness Boundaries on National Forest Lands in the Cascade Mountains, 1950–1984* (dissertation, Washington State University, 2002) adds scholarly depth to wilderness history in this region and will be published in book form.

The legislative struggle for new wilderness areas, national parks and wildlife refuges, and wild and scenic rivers across Alaska is chronicled by Daniel Nelson in *Northern Landscapes: The Struggle for Wilderness Alaska* (Washington, D.C.: RFF Press, 2004).

Wilderness Stewardship and Wilderness Values

On stewardship, there is but one book to list, for it is both excellent and comprehensive: John C. Hendee and Chad P. Dawson, *Wilderness Management: Stewardship and Protection of Resources and Values*, 3rd ed. (Golden, Colo.: The Wild Foundation and Fulcrum Publishing, 2002). This is, quite literally, the bible on the subject.

Another bible, this bringing together the latest insights on each of the many values and benefits of preserving wilderness, is H. Ken Cordell, John C. Bergstrom, and J. M. Bowker, eds., *The Multiple Values of Wilderness* (State College, Penn.: Venture Publishing, forthcoming in 2005). I have a chapter in the book summarizing the Wilderness Act and the history of its implementation.

Author's Note and Acknowledgments

Author's Note

While this is a Campaign for America's Wilderness book, the opinions expressed are my own.

In delving into the history introduced in this volume, I do not pretend to be a dispassionate historian. Where I was personally involved, I have used a first-person idiom. Participant-written history is the most suspect kind, tempting the writer to burnish his or her own role and settle old scores. I have sought to be a respectful student of this history, mining it for lessons I have found extraordinarily useful in my own advocacy and in helping others.

Acknowledgments

Longtime friends and colleagues in this work, John McComb, Chuck Clusen, Ed Zahniser, Tim Mahoney, Joe Walicki, Buck Parker, Andy Wiessner, Bruce Hamilton, Bart Koehler, and Brock Evans, have each deepened my understanding of wilderness politics more than they can know. Jim, Sandy, Larry, Patti, Consuelo, and James have kept my head above water when I needed it most.

I am indebted to Ed Zahniser, who literally grew up with the Wilderness Bill. John McComb has been a source of friendship and wisdom; his unsung strategic talents are of the highest caliber. John, with Brian O'Donnell, Chris Barns, and John Gilroy, diligently commented on parts of the manuscript.

I have been blessed to work at the side of extraordinary wilderness leaders since very soon after the Wilderness Act became law: Stewart M. Brandborg, Rupert Cutler, George Marshall, Harry Crandell, Ernie Dickerman, Mike Nadel, Dick Olson, Dave Foreman, and Clif Merritt of The Wilderness Society; Mike McCloskey, Lloyd Tupling, Edgar Wayburn, Carl Pope, Jonathan Ela, Dick Fiddler, Jack Hession, Brock Evans, Jim Blomquist, and so many others in my Sierra Club family; Dave Brower and George Alderson of Friends of the Earth, and Cindy Shogan of the Arctic Wilderness League.

Outstanding colleagues at the state level deepened my understanding of grassroots organizing, exemplified by Larry Williams (Oregon), Rick Johnson (Idaho), Doris Milner (Montana), Polly Dyer and Don Parks (Washington state), Jim Eaton (California), and—as examples of younger stars—Heather Morijah (South Dakota) and Kirk Johnson (Pennsylvania). Naming some in no way is meant to slight hundreds like them.

My special thanks go to Mike Matz, John Gilroy, Susan Whitmore, Ken Rait, and each of our colleagues of the Campaign for America's Wilderness who teach me something new daily; to Jane Danowitz of the Pew Charitable Trusts, and to Bill Meadows of The Wilderness Society. Bart Koehler and his colleagues at their Wilderness Support Center are an invaluable resource for wilderness, if the travel doesn't kill them.

•

Many historians have tolerated me loitering around the outer defenses of their profession. I am grateful to Roderick Nash, Dennis Roth, and Gerald Williams of the Department of Agriculture, Richard West Sellars of the National Park Service, Mark W. T. Harvey, John Miles, and James Morton Turner. Devoted researchers H. Ken Cordell of Forest Service Research and Peter Landres of the Aldo Leopold Wilderness Institute and their colleagues are ever patient with my requests. Thanks, too, to Ann Lage of the Bancroft Library, University of California, Berkeley, Carol Winn of the University of Washington, and Barbara Walton, formerly with the Western History Collection at the Denver Public Library.

The giants of the Congress with whom I have been so lucky to work with for wilderness preservation are extolled in this volume—Saylor, Church, Udall, Seiberling, Burton—and they are simply first among many as each would have been the first to say. As I began my career, four legislators from Michigan made me a lifelong believer in the power of our Congress for good—Representatives Marvin Esch, Guy Vander Jagt, and John Dingell, and particularly the most saintly and self-effacing politician I have ever known, Senator Philip A. Hart.

Connie Myers, founding director of the Arthur Carhart National Wilderness Training Center, and her colleagues are deeply devoted to a wilderness-forever future; as they train, they share their passion for the wild with thousands of agency wilderness stewards and leaders each year. There are unheralded champions of wilderness throughout each of the four federal agencies. However, as I write this it is the better part of discretion to thank them anonymously for their patient advice— they know who they are.

Some of the ideas and formulations found here first appeared (or will soon) in *Wild Earth*, the *International Journal of Wilderness*, *Wilderness*, and *Sierra*, as well as in publications of the Campaign for America's Wilderness.

No author of a first book has been blessed with better partners than the people of Fulcrum Publishing. Vance Martin, a colleague of theirs who directs the WILD Foundation, came to me with the idea of this book. Sam Scinta and Kristen Foehner embraced the book with tireless enthusiasm. No writer stuck in the deep ruts of his own styling and imaginative spellings can have a more devoted editor than Faith Marcovecchio.

Through all of this history—and the history of wilderness preservation still being written—the greatest achievement is that of tens of thousands of spirited citizen-activists who, as volunteers, have indeed fought, in Bob Marshall's immortal phrase, "for the freedom of the wilderness." Our National Wilderness Preservation System is, above all else, *their* gift—and Howard Zahniser's—to the Earth and to future generations.

—Doug Scott
July 2004

About the Author

(Courtesy Kiana Scott)

Doug Scott is policy director of the Campaign for America's Wilderness, working nationally from his office in Seattle. He holds a forestry degree from the University of Michigan (1966), where he did his graduate research on the history and drafting of what became the Wilderness Act of 1964.

Mr. Scott began his own work for wilderness preservation soon after the Wilderness Act became law. As a volunteer activist while in graduate school, a Washington lobbyist for The Wilderness Society, and northwest representative for the Sierra Club, he was in the forefront of many of the important wilderness preservation campaigns as a strategist and lobbyist. These include enactment of the Eastern Wilderness Areas Act and Hells Canyon National Recreation and Wilderness Act (both 1975), the Alpine Lakes Wilderness Act (1976), the Endangered American Wilderness Act (1978), the Central Idaho Wilderness Act and monumental Alaska National Interest Lands Conservation Act (both 1980), statewide wilderness laws for dozens of states in the 1980s, and the California Desert Protection Act (1994).

In the 1980s Mr. Scott was conservation director and later associate executive director of the Sierra Club. In 1996 he received the club's highest honor, the John Muir Award.

Mr. Scott speaks frequently on college campuses, at training programs for federal agency personnel who administer wilderness areas, and at gatherings of wilderness advocacy organizations. He was a member of the board of directors of Environmental Teach-In, Inc., which organized the first Earth Day in 1970. He has served on the board of directors of the League of Conservation Voters and People for Puget Sound, and is currently on the board of The Wilderness Land Trust.

CAMPAIGN
for AMERICA'S WILDERNESS

All the wilderness future generations of Americans will ever have the chance to know is that which our generation protects today with the full force of federal law.

The Campaign for America's Wilderness works to translate broad public support for wilderness preservation into action that can propel citizen wilderness proposals to enactment by Congress. Our efforts are positive and proactive. We collaborate with state coalition and local citizen groups as partners, aiding them with additional resources and expertise—media, organizing, policy, and advocacy for wilderness in Congress. For further information or to get our monthly electronic newsletter, visit www.leaveitwild.org

Grassroots Assistance:
Jen Schmidt   (212) 645-9880 x17 jschmidt@leaveitwild.org

Media and Communications:
Susan Whitmore   (202) 266-0435   swhitmore@leaveitwild.org

Policy:
Doug Scott   (206) 200-0804   dscott@leaveitwild.org

The Wilderness Support Center

We are close partners with our colleagues of The Wilderness Society's Wilderness Support Center, who do this same work. By each taking the lead in assisting partners in specific states and by drawing on the unique expertise of each other's staffs—as well as many other national, regional, and state organizations—we all strive to promote the strongest, most effective wilderness movement for America.

Contact: (970) 247-8788   wsc@tws.org

http://www.wilderness.org/OurIssues/Wilderness/wsc.cfm

Speaker's Corner Books

A provocative new series designed to stimulate, educate, and foster discussion on significant public policy topics.

Parting Shots from My Brittle Bow

Reflections on American Politics and Life

> Eugene McCarthy
> ISBN 1-55591-528-0
> 6 x 9
> 160 pages
> $12.95 pb

A contemplation of the tides of American politics

The Brave New World of Health Care

What Every American Needs to Know About Our Impending Health Care Crisis

> Richard D. Lamm
> ISBN 1-55591-510-8
> 6 x 9
> 132 pages
> $12.95 pb

An exposé of health care policy in the United States

Fulcrum Publishing
16100 Table Mountain Parkway, Suite 300, Golden, CO, 80403
To order call 800-992-2908 or visit www.fulcrum-books.com.
Also available at your local bookstore.